# PHANTOMS OF THE PRAIRIES

**Other Books by**
**TANNER LAINE**

**Campfire Stories**
**Cow Country**
**What's In A Name?**

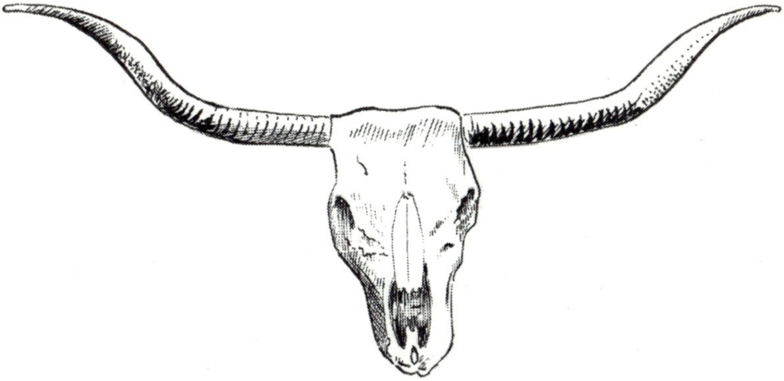

# Phantoms of the Prairies

By Tanner Laine

PIONEER BOOK PUBLISHERS, INC.
P.O. Box 1003
Hereford, Texas 79045

Printed in the U.S.A.

# FOREWORD

**STOP!**

**You are in the land of the wide sky and beckoning horizons. You are in West Texas.**

**LOOK!**

**Is that a wisp of white as the shy antelope turns tail to your dared intrusion?**

**LISTEN!**

**Can you hear the far off low thunder of the buffalo herd? The faster, sharp, "clack-clack-clackedy-clack" of many mustang hooves! Perhaps even, the dry whirling buzz of a rattlesnake's tail.**

**Many ghost herds, animals and birds, flitter again on these pages of "Phantoms of the Prairies", as they do in memory for the many of us.**

**That is the purpose, then, of this book—to touch ever so lightly the "Lost Chord of Memory".**

**Of course, many of the "phantoms"—so called here—are not extinct or a long-lost breed. It's just that urban living of today has pushed them so far back in remote pastures and canyons. It is reluctantly true: The bitter sweet songs of the mockingbird or meadowlark never sound as good in a city-surrounding as they do along the lanes of the rural areas of our childhood.**

**—Tanner Laine**

## PHANTOMS OF THE PRAIRIES
## Table of Contents



***Continued***

# # #

## THE THUNDERING HERD

As the old hunter said, "It took a lot of gun to kill a buffalo.

He was, of course, referring to firepower of one weapon.

So it did. Perhaps the reason J. Wright Mooar was so successful (he killed an estimated 20,000 buffaloes) was that his rifles were hefty 50-caliber jobs.

Even the Comanches stood in awe!

It could be they were watching one day when J. Wright Mooar made one of his greatest shots.

He was near the present Lynn-Garza counties line when he saw a giant buffalo bull.

He sighted in on the animal. He was afraid he would be unable to approach nearer to get a better shot.

His "Big Fifty" roared! The bull fell. Mooar stepped off the distance of his shot, and found it to be 756 steps.

On Oct. 7, 1876, two significant events occurred. One, the Mooar hunting party arrived at Deep Creek, about 10 miles northwest of present-day Snyder. And, on this day, several hundred yards from camp, J. Wright Mooar shot a rare white buffalo.

Records say it was one of seven albino buffalo to be seen, or killed, among all the millions of bison that once thundered in herds by the millions in pioneer days.

It is fitting that a handsome statue of a white buffalo is located on the courthouse square in the now All-American City of Snyder.

* * *

On July 5, 1972, a buffalo calf was born at the site where Lubbock is, on the South Plains.

An oldtimer joked: "Bet you it's been almost a century since that happened!"

But it did happen.

Appropriately and patriotically, the buffalo was named "Firecracker," its birth coming only a few hours after the Fourth of July.

It came about this way:

Hugh ("Hackberry") Johnson, 84-year-old western showman, stopped off at Lubbock en route to the annual Indian Tribal Ceremonial at Gallup, N.M. "Hackberry" had a trailer load of buffalo hides and four live buffalo.

He has been taking buffaloes to Gallup for eight years. "The Indians try to ride 'em," Johnson chuckled.

"Hey, I guess I'm the last of the 'hide hunters,'" Hackberry commented, referring to the once lucrative industry on the plains and prairies of picking up buffalo hides and transporting them to railheads.

"Hack" is a well known western character in West Texas. He got his name after losing a leg in a ranch accident when he was a cowboy.

He fashioned himself a peg leg out of a hackberry tree limb to replace the limb he lost. He is said to be one of few who competed in rodeos with a wooden leg. But he did just that.

Needless to say, the presence of buffalo in 1972 at Lubbock, gave rise to many stories of this area, when great herds of buffalo thundered by, and millions of the burly animals were grazing in the Southwest.

# # #

## OLD HAWKEYE

The settler's son ran to the house, excited and yelling: "Mom, mom, that old chicken hawk is flying around our chicken house again!"

His mother sprang to action. "He is, is he? Well, I'm going to make some feathers fly and they won't be chicken feathers," she said, as she grabbed up her husband's .22 caliber rifle.

As she slammed the back door and headed for the barns and chicken house, the gliding hawk's sharp eyes swept her and saw the glint of the rifle barrel in the sun light.

The hawk banked sharply, and flew rapidly away, its wide wings pumping, into the skies beyond rifle range.

"Old Hawkeye," as the settler family had dubbed him, had escaped again.

* * *

This was not an uncommon episode in this part of the country in pioneer, settling days, and even later.

The hawk is a killer.

He belongs to the family of eagles and is a large bird of prey.

Hawks are fairly silent, but when disturbed, utter piercing whistles, screams, and chattering calls.

Hawks capture living animals and kill them instantly for food. With swift wings, they pounce on prey with lightning speed.

The hawk's eyesight is sharp and keen. It is good from very high altitudes, particularly for moving objects on the ground.

Some of the larger kinds of hawks have been known to attack people who came close to their nests, causing painful cuts with sharp talons.

There are two main kinds of hawks in North America—bird hawks and buzzard hawks. The so-called chicken hawk is a bird hawk, so called because of a liking for poultry, especially chickens.

# # #

## THE POOR MAN'S GROUND HOG

Prairie dogs, whose towns once dotted the plains and prairies from one end to another, served as the cowhand's groundhog.

This is proved in a little story that goes something like this:

"Hey Jake, it's not long 'til warm weather and you can shuck your long handles," observed a Spur Ranch cowpoke in an early-day bunk house session in this area.

"Heck, I already know that," drawled the late Jake Raines, one of the oldtime cowhands at the foot of the Caprock.

Both cowhands had seen the same prairie dog town on a morning ride that day, and the 'dogs had been out with their young ones, nibbling a few grass roots.

Both knew this was a sign of warm weather.

Now these cowhands knew nothing about Groundhog Day as a harbinger of weather for the next six weeks, but they did have their signs of approaching spring in lots of "folk forecasts."

There weren't any weather instruments in those days but nevertheless, the cowhand could read signs in nature and animals and could forecast weather changes and season arrivals with uncanny accuracy.

* * *

Although he was well-entrenched here before man came west in West Texas, the little barking rodent got in the proverbial "doghouse" with ranchers pretty quick.

It was the economic factor. Mr. Prairie Dog lived on grass and he could destroy vast areas of grazing land. Also, cattlemen's horses sometimes stepped in his holes, breaking their legs.

The prairie dog became sort of a "Public Enemy Number One" among animals under Texas law. This law was rough on "Prairie Dog Pete," and drove him pretty far back in his maze of tunnels.

His extermination was widespread.

Legislation against the prairie dog, providing for local option on the poisoning of his kind, dates back to an act of 1903. He was declared a public nuisance and it became the duty of landowners to destroy all the prairie dogs on their land.

There also were provisions whereby commissioners courts were to investigate to see that landowners destroyed prairie dogs.

But in 1971, this old and ancient law got a trip to the "bone pile" and the prairie dog was at long last, pardoned.

* * *

Fact is, old "Pete" already had begun to show himself. Modern-day residents considered him a novelty and established towns as a tourist attraction. Lubbock has capitalized on a "Prairie Dog Town" for several years.

**And at Andrews in 1971, newspaperman James Roberts discovered prairie dogs in their towns are a lot like people. His column accounted:**

(When 28 prairie dogs were ordered by Andrews from Odessa, the newcomers were dumped into the Andrews colony, where five old prairie dogs remained. Roberts, who had heard from pioneers that prairie dogs from different towns would not mix, got out his binoculars and watched):

"Soon, there was war; rumors of war; and the isolationists cried for peace.

"The five old residents, fat and sleek from three years of prosperous living, were suddenly under pressure from 28 newcomers—lean, young and hungry.

"The clamor set up sounded like the Tokyo Stock Exchange.

"Early in the warfare, the old 'dogs acted true to 'established form'—trying to defend at least three holes each against the newcomers at one time.

"The old ones had more weight and familiarity with the town, but the young 'dogs had more speed and agility.

"The encounters were both furious and bloody. Time and time again, the old sent the young hustling away.

"But battle weary, tired and feeling their age, the five old 'dogs retreated at last under pressure—as strong as the dollar has been under in foreign markets, lately.

"The old ones, finding they couldn't bluff the young invaders, tried to dig furiously and cover up all the holes and newcomers—the whole works.

"Finally, the 'establishment' sought a corner of its own town, and from a prone position, eyed the newcomers with contempt but no physical backing.

"But, if they are not careful and the park becomes segregated, the first thing they know, HEW will insist on busing young prairie dogs from one side of the park to the other," thus ended the parable of the prairie dogs.

# # #

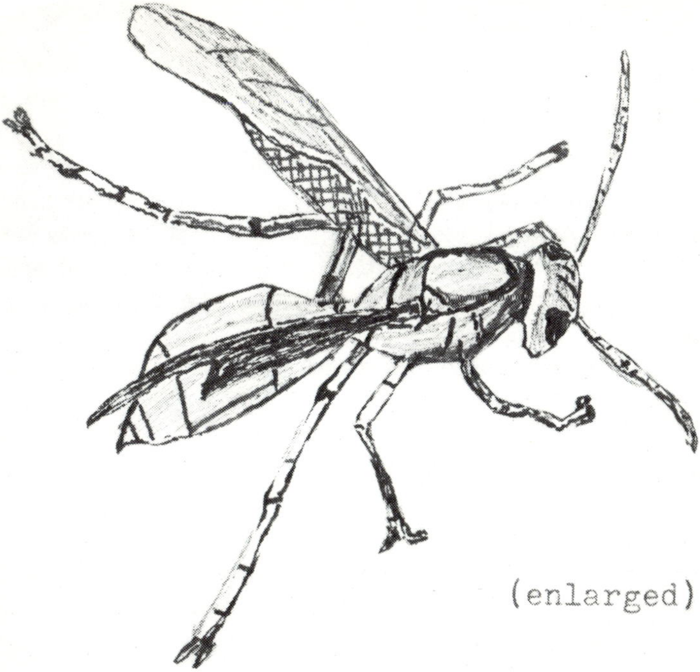

(enlarged)

## YELLOWJACKET

Some of the early-day phantoms of the plains and prairies packed quite a wallop! Where some merely frightened their victims, some could sting realistically.

Among the stingers was the "yellowjacket" (wasp). He could drill man or beast, producing an anguished screech or spirited jump, or both!

This yellow and black-marked fellow was in abundance on ranches and farms. He hung his multicelled nest from bushes and trees and under rafters. He ate the nectar and juices of ripe fruit.

* * *

Mrs. Mary Stovall Lambert of Munument, N.M., writes in some of her "Experiences As A Bride On The Pitchfork Ranch" of her encounter with this stinging kind of phantom:

"Horseback riding was my favorite pastime when I lived on The Forks. I rode many miles over this spread of ranchland in the "Big Ranch Country," east of Dickens.

"One time, Mrs. Fred Haile, a friend of mine from Spur, came out to the ranch for a visit. We decided to go horseback riding.

"On returning to the ranch house after a long ride, we spotted a 'ripe plum thicket.'

"We rode right into it to pick some of the enticing sun-red ripened fruit.

"Just as I reached out for a big juicy plum, a disturbed 'yellowjacket' hit my pony, 'White Eagle.'

"White Eagle forsook his gentle ways and let down none too easy! I went sailing off into that thorny thicket.

"I know that sting must have smarted plenty, or else 'White Eagle' never would have thrown me.

"I got right back on my horse after I had settled him, and we rode on home.

"Our appetite for plums was suddenly gone for that particular day."

# # #

## MOLLIE'S SECRET

The old cowhand has to explain to a visitor from the city that the nests of dried branches found in remote parts of the pastures are not hastily-built campfires that were never lighted.

These, in truth, are nests of wood rats that got their name from this practice. The "rat's nests" are made of assorted-size limbs and branches in brushy country.

The wood rat lives in the Southern and Western parts of the nation.

His collection of materials did provide starting "fixings" for a quick branding iron fire or campfire since the materials he could handle were fairly small.

The finding of a wood rat nest in the fireplace of a tattered, empty old ranch or farm house—as if a fire had been laid—is one of the amusing moments of visits to these old houses in West Texas.

Today, you very seldom see a wood rat nest, and then, only on the rarest of occasions, do you see the maker of such a structure.

* * *

Another name for the wood rat is "pack rat."

As a lad, we had learned this and therefore every nest was searched for something "shiny" as being possibly of value.

The wood or "pack" rat is curious. It will pick up small articles that catch its fancy, such as coins, belt buckles, nails, etc., even bright-colored stones.

For this reason, the name of "pack" rat is applied because the rodent packs the objects to its nest.

We never found anything of value that had been "packed" by a rat, although there were stories of lost rings and old coins being found in rat nests.

And that brings to mind a little story:

* * *

Mollie Vest was her name. She was a belle of the ranch country around Midland. She was a pretty Irish rose in a prairie land. She was bound to have been, pretty as she was at 80 plus.

Her suitors were many among the cowboys and settlers. Mollie was pert, as she was pretty. She was impetuous and wore an impish smile.

Small wonder that one cowhand suitor pressed his courting case emphatically. He even bought a band of gold and roughly put it on Mollie's finger.

But she slapped his face, and tossed the ring into the night, somewhere in the rugged rangelands around the couple.

Wouldn't it have been something had a wandering pack rat found it, and scurried to a nest with the bright band of gold?

Even though it might be found now, say by an oilfield scouting and staking crew on location in the oil-rich Permian Basin fields, the finder would never know the secret of the band of gold and how it got there, way out on the prairies.

For you see, Mollie loved Frank McCormick, another suitor than the impatient one that tried to give Mollie the ring. And it was Frank she married happily. Both now are sleeping in Midland's Fairview Cemetery—and Mollie's secret will be...forever.

# # #

## SCISSOR-TAIL

The cruel chicken hawk soared, then swooped low over the brush-strewn ranchland at the foot of the Caprock.

Spotting a family of bob white quail, the big bird gained altitude, then slanted off into a steepening glide toward his intended prey.

But it was not to be.

For suddenly, a pair of trim birds with long tails swept in to buzz the predator.

Pressing a relentless attack, able to make tighter turns, and possessing greater flying skill, the smaller birds soon, literally, put the hawk "to flight"—away from the intended prey.

They had attacked from behind, driving their beaks at the back of the hawk's head; then, circling, had flailed him with their long tail feathers, which opened and closed like a pair of scissors.

Anyone who has ever witnessed this bit of aerial warfare above the prairies of West Texas, cannot fail to have admiration for the scissor-tail.

His aerial attacks on the hawk resemble the onslaught of small fighter planes against a larger bomber.

Scissor-tails prefer remote rangelands in which to live and nest. They have been a big favorite in West Texas for a long, long time.

About the size of a mockingbird, the scissor-tail resembles the singer in color and marking, except for the distinctive tail.

The scissor-tail is a flycatcher, but rarely is identified as such out here. Preferring warm climates, the scissor-tailed flycatcher is one of the most striking birds of the Southern High Plains, New Mexico and Western Texas.

What a fearless scrapper he is!

# # #

## PULLING THE BADGER

The badger is an animal of the same family as skunks and weasels. He is about 2 feet long with a flat head and short neck. Its legs are thick and strong and the four feet have long claws, used for digging. There is a white stripe running along the back. The body is wide and flat.

The badger digs its hole on level prairies, or in woods, where trees are far apart. He eats field mice, gophers and ground squirrels.

Mr. Badger is quick-tempered and quarrelsome. He is brave and will fight savagely. Often the badger was matched in a fight with dogs, which were pushed into sort of an arena, where a badger would be in a barrel. And the dogs were to "pull him out." This is said to be the source of the expression "to badger" or tease.

The badger's presence out here in West Texas gave rise to a popular practical joke that was played on many a "tenderfoot" or "greenhorn" or just plain stranger to these parts.

This was variously called a "badger fight" or "badger pull," etc.

* * *

You see out here in West Texas in the early days, entertainment was pretty scarce. A practical joke that the residents of the small towns could enjoy and talk about for days, was appreciated.

There arose this type of entertainment that was perpetuated in almost every town in the Plains area. It was the "badger fight" or "badger pull."

Now, badgers were real animals out here. And if you ever saw one, you could easily doubt that any opponent of comparable size could stand up against the powerful, squat animal in a fight.

Let some stranger come around and immediately the local boys would start "talking up" a badger fight.

Some rancher or farmer or town resident always had a dog that he loudly swore could whip any badger that could be found.

The argument would begin. It grew hotter and hotter until finally, money was put up and a fight would be arranged.

The stranger, as an unbiased party, would be asked to "pull the badger"—that is release him from under some box or barrel when the fight was to come off.

And brother, what that unsuspecting stranger pulled out would be fodder for laughs for days to come.

* * *

Ross Edwards, an oldtimer living at Lubbock, can tell about one he witnessed and wrote about in his book "Fiddle Dust":

". . . to rid the country of the pest prairie dog, the Swensons decided to go the poison route. They equipped two wagons with men to put out the poison. . .

"While this crusade was on, a couple of young men from New York, who were interested in securing Western lore for a book, came to Spur.

"These two donned overalls and went out with the 'dog poisoning' crews. The outfits always came to town on Saturdays. One day, they caught some real badgers out on the range and brought them to town.

"This revived an old custom that had been practiced for many years...that of having a dog and a badger fight. Everyone put up a great show of interest and spread the news up and down the street...After a few weeks, the Saturday arrived for the fight and the two New Yorkers were to pull the badger.

"Well, we arranged for the fight to be pulled off back of Lowery Drug Store. Bryant-Link Co. had a two-story building that extended quite a bit back of the drug store. This would enable a sizeable bunch of ladies to view the exhibition, usually reserved for the more immodest sex.

"A large box was placed on the ground in advance to "cover" the badger. This pseudo 'badger' was a traditional utensil that reposed under the bed in a day when there were no bathrooms. For this particular fight, a rope was attached to the handle of the 'thunder mug' and stretched out to a convenient length to the puller.

"The supposition is that a great advantage can be obtained for the dog in the manner by which the badger is 'pulled.'

"Therefore, someone who would be absolutely fair and not betting on the dog or badger, would be the one and only one agreeable to both sides of the betting ring.

"There was a lot of money waved around. There was so much hub-bub that I almost thought the 'frame up' was real.

"The New Yorkers were persuaded to pull the rope. Everyone agreed they would be just the ones to do it.

"Some impious individual had poured two bottles of 'Coke' and put a box of ginger snaps into said 'mug.'

"Dr. Bachelor would lead his old bird dog around and when he got up to the box, that old dog would set up a tremendous barking.

"The New Yorkers were to pull in unison. They got set and at the crack of a pistol, yanked out the badger, onto themselves and anybody up close, sloshed the entire contents.

"After the jubilation was over, we all shook hands. I wish I could remember those boys' names. The old West never had two better sports..."

# # #

## WHO! WHO?

Owls made good phantoms, winging their way in the dark shadows of night on early-day ranches in West Texas; or, sitting on a tree, their yellow eyes shining, and giving out with a haunting "Who! who! who are you?"

Mrs. Orwin Lambert (Mary Stovall), Monument, N.M., has a good story about an owl. She resided on the Pitchfork Ranch in West Texas, after she married Orin (Red Mud) Lambert, and went to the big ranch as his bride:

* * *

"One real cold night about midnight, we heard the chickens being disturbed.

"Orwin got up and went to investigate.

"Sure enough, there was an old owl, perched up in a tree.

"Orwin shot him down and brought him by the house to show me. Then he showed it to the boys at the bunkhouse.

"This done, he just threw the old hooter down.

"Next day, after the boys had ridden out to work, I went out to see the owl but couldn't find him anywhere.

"Finally, I found him behind the bunkhouse, and he wasn't dead.

"I carefully caught him and took him to the house, tied up a leg, and fed him.

"When the boys came riding in, I told them I wanted to show them an owl that I had caught. They all trooped in and looked him over.

"Shucks, my husband declared 'I shot one twice that big last night right here.'

"How big?" I asked Red Mud.

"Twice as big as that one there,' he declared.

"'That's funny,' I laughed. 'This is the same old owl that you shot.'

"I am afraid that my husband didn't appreciate the prank I had played on him, especially with all those cowhands ribbing him like they did."

# # #

## WALK 'EM DOWN

Mustangs—wild and free as the wind—ran unshod over the entire Southwest in an earlier day and chapter of West Texas history.

These horses were descendants of the horse stock brought into Mexico from Spain by the early Spanish conquerors.

Then, as Spanish explorers made numerous expeditions, some of their horses, no doubt, escaped in time of storm, buffalo stampede, or on staggering marches to find water and sufficient grass.

Scattered this way in the Plains country, these beautiful little inbred ponies were found on the western ranges by the hundreds.

These were true mustangs, named by inhabitants of Mexico.

When the wagon roads were cut across the Plains to California and to the various military posts of the Southwest, horses sometimes escaped and joined the bands of mustangs.

Horses, stallions and mares, escaping from wagon trains across the Plains, intermingling with pure-bred mustangs, caused the native mustangs to become more and more scarce. By the late 1800's, almost all of the pure mustangs had disappeared.

Today, a few are found among Indian pony herds on some reservations, where the breeding-up process to get larger horses to haul freight and till the soil, has not been rigidly enforced.

* * *

More about the earlier mustangs: their average weight was 800 pounds. The colors were cream, buckskin and mouse-color. Some had black stripes about the legs and on the back. Manes were rather large and surprisingly, did not have a shaggy appearance.

They were clean-limbed and their hoofs were black and perfect. They had never tasted grain. They certainly were hardy. Broken to ride, they could stand plenty of hard riding.

According to the late Joe Seagler, historian and researcher, blacksmiths or "hoof-shapers" never had to tinker with mustang hoofs, or force the animals to wear iron shoes, because the strong hoofs could stand wear over the roughest of terrains.

* * *

There were many ways to capture a wild mustang.

Hale County pioneers Thornton and Will Jones had a unique way:

Catching mustangs, they would walk them down! They said if you could keep the mustangs awake the last two hours of the night for about a week, they would get so tired and sleepy you could get close to them.

They would get a bunch of eight or ten walked down, then take a big strong saddle horse and 100 feet of rope, tie one end to the saddle horn and at the other end, make a loop, and sit down by the horse.

Another man would ride around the wild horses until they were over the loop. When one stepped in the small loop, the man by the horse would jerk and catch a wild horse by the forefoot.

Oh yes, the mustang would pitch and buck but would wear down. Two men could get other horses from the wild band, because they wouldn't run far after one was caught.

Then the wild ones were thrown in with gentle horses and driven to a corral.

* * *

The late Max Coleman of Lubbock, a mustanger, once told how he found a Mexican on the Gutierez Ranch of Old Mexico, who could flip a rope in half circle that would run down the rope, looping around the horse's nose. This was the half-hitch. Coleman declared he could never throw a half-hitch from a moving horse himself and had only seen it done once.

# # #

## THE ALMOST BLIND TUNNELER

On occasions, some early settlers of this part of Texas, in early morning, found a maze of burrows and tunnels spoiling their gardens.

This was laid to the mole.

The mole you seldom see anymore is a small, thick animal that is a tireless and rapid digger.

Moles can dig their way through about 300 feet of earth in one night, as they search endlessly for earthworms, which are their chief food.

He is always hungry and eats his own weight in food in a single day. If the mole has no food for 10 to 12 hours, it will die.

Moles are shaped by their burrowing life underground, which is like a plow. The common mole is only five or six inches long. It has a narrow, pointed nose, wedgeshaped head, and enlarged forelegs. These legs work like a chisel to gouge out the earth. The hind legs are short and powerful.

Moles have pinpoint-sized eyes, which are shaded by an overhanging fur. Their underground life makes it unnecessary for them to see, and they are almost blind. Their small ears have acute sense of hearing.

These animals burrow mostly at night, and prefer to stay inside their underground fortress homes during the day.

The fur of moles is valuable. Moleskin coats and jackets are expensive. Perhaps this has contributed to the decrease in these animals. The handsome and soft, pleasing fur of a mole makes a fine evening wrap.

The common mole is found in almost every part of North America, as far south as the Mexican border. Moles also are found in Europe and Asia.

# # #

## LONGHORN

In 1493, Christopher Columbus left a few Spanish cattle in Santa Domingo. In 1521, Gregorio de Villalobos took some of this stock to Mexico. Other importations followed, all of them potent to the breed—we call Longhorn Cattle.

The Spanish took along cattle when they established missions and colonies. Spanish settlements were concentrated along the San Antonio River in the 18th Century.

By the time Stephen F. Austin's colonists reached Texas in the early 1820's, cattle of the Spanish type ran wild from Red River to the Rio Grande.

Some of this blood went into the Longhorn type that became synonymous with Texas. But the Texas Longhorn was a rangier breed. Into this breed filtered some of the Shorthorn cattle. The range cow was beginning to take shape—hoof and leg designed to walk long trails; ability to rustle for themselves; ability in drought to go miles without water; and if there was no grass, to exist on brush and prickly pear.

Trail driving began in the 1860's. During the 25 years following the Civil War, it is estimated 10 million Texas Longhorns were trail-driven to Mid-west feeders and Kansas railheads.

But the Longhorn was too much bone, leg, flank and horn, in proportion to beef. As land increased in value, more valuable beef cattle were required. Barbed wire was introduced. Breeding programs were pushed.

The Longhorn had to go...

* * *

Red-eyed Longhorns could be pretty scary in the early days of the Plains country, like a charging, mean one told about in a true story related by Mrs. C.L. Simmons of Plainview:

"If my father were living today, he would say it was a 'miracle' that he lived through that experience.

"In 1890, Papa sold his little farm in Wise County and started out West, with his family, to seek his fortune.

"This new country looked quite different from Wise County, where most of the pastureland was dotted with post oak and blackjack trees.

"Where we located, the pastures were all prairie except for a few scattered mesquite bushes.

"Little did Papa realize that he would be wishing for one of those old post oak trees, and soon.

"One day, Papa was plowing nearly a mile from the house, and old Pete's bridle broke. Leaving the horses to 'blow a spell,' Papa started to the house to mend Pete's bridle.

"Taking a short cut across a pasture, he saw a herd of Longhorns grazing peacefully enough—all except one old critter that was eyeing Pa, suspiciously.

"That one, a bull, became angry at seeing a human crossing his domain. He began pawing the earth, then shook his head of wide horns at the intruder.

"Papa hurried his steps. Glancing back, he saw the old bull was angling toward him, and then it charged!

"That was when Papa missed his post oak trees! How he longed for one to climb. The faster Papa ran, the nearer that old Longhorn came! Just as it looked as if he were a goner, Papa had a bright idea.

"Turning to face the raging animal, he threw the bridle smack into its face. It caught on one of the horns, obscuring view from one eye.

"With the other, eye, the bull watched his fast-moving prey get away.

"A few days later, old Pete's bridle was found hanging on a mesquite bush near where Mr. Longhorn had stopped short of goring Papa."

* * *

Mrs. Leona Sparks Koonsman of Dickens writes that her husband, Jim, was one of the ropers that cleaned out a herd of rangy and mean wild steers, Longhorn variety, that lurked in "Lost Canyon" of the Croton Breaks:

"The cowboys roped and sawed the horns off of the big steers that had strayed from the Matador Ranch and the Spur Ranch (later Swenson Ranch). Tud Arthur was camp man at Croton Camp at the time. He, Jim Koonsman, Cliff Bird and Edd E. Smith would go out on their horses and rope the big steers, heel them, and then saw the horns off. One of the sets was put on display in the Dickens County Museum. Many stories about these steers—phantom and real—persisted for many years. No more steers like those can be found anywhere today.

"Some say, when the night is stormy and the wind is whistling, these old steers come out of the canyon once again and cut up out in the breaks—but I have never seen one!"

# # #

## "PHEW!" SNIFFED "BR'ER" BEAR

"Bears in West Texas?"

"Preposterous!" says just about everybody today.

But the late W.L. (Uncle Bill) Hyatt, an oldtime cowpoke of the Spur-Dickens-Matador area, once told this story:

"I came to the edge of the Caprock and Plains country in 1878, with a herd of the old "22 Ranch" outfit.

"This was only a year or so after the big Indian fights (Mackenzie's Raiders vs. the Comanches). And a few renegade Indians still roamed these parts."

"There was no one out here except a few Indians, and buffalo hunters. The cottonwood limbs along creeks and hackberry groves in the draws, were full of wild turkeys.

"Antelope timidly browsed and grazed. There were wild mustangs, panthers and rattlesnakes.

"There were bears, too, at least one. I saw him!

"One night I was standing guard for Indians. I saw something stealing up on us.

"We always slept in a circle, like spokes of a wagon sheet, heads to the center and feet out," chuckled Uncle Bill.

"I gave the boss a gentle kick on the foot to arouse him.

"He and I watched this thing approach.

"It was a big bear! (Probably a Cinnamon bear).

"We watched it prowl, eat some scraps the boys had scraped from tin supper plates.

"Old Brother Bear finally sniffed at one of the sleeping cowpoke's feet—and ran plumb away—into the night!

"I never saw that bear again, or any other one out here."

# # #

## 'BEEP,' THE ROADRUNNER

You may have seen him—heisting his long tail, outstretching his long neck, and come darting in front of your car, challenging for a short race, and then go legging it across the bar ditch and into the brush at a rapid clip on his long legs.

This is the roadrunner.

He also goes by the name of chaparral; or "paisano," as the Mexicans call him.

Whatever name he goes by, he is one of the most conspicuous and intriguing birds of the Southwest.

Of the 825 kinds of birds in Texas, the chaparral is a favorite of many, and has been since early days. The "paisano" is the state bird of New Mexico.

His habit of running on the ground, seldom flying, although he can hedge-hop to escape his enemies; and his prowess in fighting rattlesnakes, have made him, almost a legendary character in folklore.

The roadrunner's speed on foot really is remarkable, although this often has been somewhat exaggerated.

He dines on lizards, grasshoppers, small snakes, insects, and even some wild berries.

He got the chaparral name from the chaparral shrub or bush, which grows in semi-arid country.

The cowboys liked him. If they had time, they had a little fun chasing the big bird on their cow ponies. They respected him for his fabled courage in fighting rattlesnakes, so they seldom killed a roadrunner.

He is a large bird, and he is wild enough. Few have been domesticated as a pet. He belongs to the cuckoo family, and some call him the ground cuckoo.

Lately, the old boy has been showing up in advertisements and commercials; in children's books; and cartoons.

For a long time, the awkward-appearing and drab-plumed roadrunner was scorned by artists, painters and carvers. But not any more! He shows up in paintings, in wood carvings, in drift wood creations, and in various metal creations.

So there, old high-stepper, Mr. Roadrunner, you finally made it—into art and the high esteem of those human beings who have given you so much trouble by just daring to walk, or drive, into your native habitat.

* * *

Mrs. Ozena Norris of Route 2, Floydada, writes:

"...I'm truly sorry you hadn't the opportunity to 'meet' our pet roadrunner. We called him 'Beep.'

"We took him from a nest before he was feathered very well and his home was a chicken coop until he was old enough to be agile.

"We fed him gobs of raw hamburger meat, along with grasshoppers. We fed him and held him close, rubbing him and talking all the while.

"After the second day, he came to meet us. Later, he had free run of our place. His favorite spot was the kitchen window. He'd sit there and watch us. His favorite roosting place was a storage closet in our garage...We could rap on that wooden coop, and call 'come on, Beep,' and he'd come running.

"Often I went outside and called him to me. If anything excited 'Beep,' his topknot feathers stood straight up and fiery red skin was revealed behind each eye.

"He would show off, if contented and pleased, by making a clucking sound and then he would do a sort of dance around me.

"He got the fever to roam, I suppose. And at a neighbor's home, a stray dog caught him. Needless to say, we missed 'Beep,' and some day hope to have another one..."

* * *

The roadrunner catches the eye of Dr. Berlie J. Fallon, of Texas Tech University. He writes:

"...Pertaining to the paisano bird, I have a collection titled 'Bird Symbols of the Fifty States'—the paisano represents New Mexico.

"In Stanza Five, I refer to a habit of the roadrunner not generally known. They lay their eggs one or two at a time over an extended period, and the first hatchings thus incubate the other birds, while the parents continue their hunting.

"J. Frank Dobie wrote an article in 1939 in which he expressed puzzlement concerning the fact that most nests he saw had a runt or two in them. At that time, he was not aware of the erratic laying habits of the paisano.

Dr. Fallon ends his verse on "Paisano Bird" thusly:

> "A place in history's fab'd lore
> Reserv'd for thee, O Noble Bird—
> In tales of taming of the West,
> The name 'Paisano' shall
> Be heard."

# # #

## THE WALKING HOUSE TRAILER

Breathes there a West Texas boy or girl who never has been fascinated by a terrapin?

Here is a crawling creature that carries his house on his back.

He can draw his head, feet and tail, into his shell house in case of danger.

He has no teeth (no turtle does) but if he snaps hold of you—wow! As a boy, we heard that he wouldn't turn loose until it thundered!

The old dryland terrapin has been around a long time. His plodding, ambling gait made the "Tortoise and the Hare" story seem even more incredible.

West Texans have painted his horny back, fed him June bugs, carved the date or initials on his back, painted or branded him—then turned Br'er Terrapin loose to roam at will.

Then looked for his return in a year, or so, and sometimes sure enough, he did come back. The markings would be there on his shell.

Turtles are interesting. They live longer than any backboned animal. There have been turtles for 175 million years. During this time, dinosaurs and giant reptiles have come and gone.

The terrapin is a box turtle. His water-favoring cousin with a softer shell, has frustrated many a fisherman out here by grabbing a bait meant for a catfish or perch.

* * *

In the 1970s, Ben Overstreet, who lives on a farm between Spur and Dickens, originated "Terrapin Races," held annually.

Ben invites the public for free to watch the races. He puts about 20 dryland terrapins under a tub in the center of a 60-foot circle. The tub is lifted and the race is on!

Spectators can buy chances or number of racers and win prizes with various winners.

How about that? Wonder what the course record is, down on the Overstreet farm?

# # #

## THE INVASION

The pesky crow is a smart one.

Crows are familiar in nearly all parts of the world.

There are many sayings about them, such as "black as a crow," etc. Since crows usually fly in a straight course, we say "as the crow flies."

Farmers out here do not like crows because they pull up sprouting grain and eat it.

Although crows have been "thinned out" considerably, nevertheless, there was an invasion by crows at Spur in 1970.

* * *

W.S. (Steve) Carlisle, who has lived in the Spur area since 1919, summed up the situation with the comment: "I've never seen anything like it. There must have been a million of them. They ate up two patches of maize of mine before we could cut it."

Carlisle's place is three miles southeast of the town of Spur.

Loyd Roberson, who has farmed two miles from Spur for more than 25 years, said crows completely ruined a 54-acre patch of wheat on his place. He said the large, black birds pulled up the young wheat sprouts before they could mature.

He said crows devoured his maize crop before he could harvest it.

"These crows were mean," Roberson said, "they literally 'pulled' most of my cotton crop—just pulled the lint and seed out of the bolls for the heck of it. I guess they were after the seed, but more often than not, they tore the cotton out and left the seeds in it."

Roy Arrington, longtime Dickens County farmer, conservatively put it this way: "Well, there were lots of crows. As bad as I have ever seen them. I cleared out the old mesquite trees where they roosted one time and they didn't come back until the young mesquite got large enough to hold them. Now they are back in force, in 1970, as bad as I've ever seen them."

Arrington came to Dickens County in 1901 and has lived at Spur since it became a town in 1910-11.

JuDon Rickels of Spur, who likes to hunt, said: "I couldn't make a dent in them. I believe there were 500,000 of them. I went out daily and shot up three boxes of shotgun shells, killing them by the dozens, and the next day it looked like more came than the day before."

Bill Gregory, vocational agriculture teacher at Spur High School, acknowledged that, "The crows were a real problem that year. They came in to roost from 5:15 to 6:15 p.m. They came from the fields around the town by the thousands. They roosted in the thick mesquite groves south of the town. Some went to roost in shade trees of Spur residents, and these trees had to be cut down to get rid of them.

"The crows got so bad a farmer offered to pay a bounty of five cents per bird to my agriculture boys to kill with clubs as they were at roost at night, but I was afraid a boy might get hurt," Gregory added.

The crows appeared as black clouds in the sky when they hovered. Waves of them approached the town, appearing as bomber formations did in World War II.

The invaders set up a loud "caw, caw, caw," if disturbed.

Several residents of the town hunted them and shot them almost daily but more always seemed to appear.

No one had a ready explanation for the larger numbers that year. And no one was quite sure how to deal with them.

"They started coming in about Thanksgiving and were at it until February and later," said a Spur resident.

The area where the crows took roost in the town was just south of W. Harris St., and south of Experiment Station Road.

Some farmers and ranchers bull-dozed mesquite brush where the crows persisted in roosting.

"They just moved over to the next thicket of mesquites," the disgruntled farmer said.

John F. Moore of Spur, member of a pioneer ranching family, who now works for the Dickens County Soil Conservation District, said you could almost set a watch by the crows twice daily—once in the morning about 7 a.m., when they took off to scour for food in the fields; and in the later afternoon just before 6 p.m., when the black formations came in for landing at the favorite roosting spots near the town.

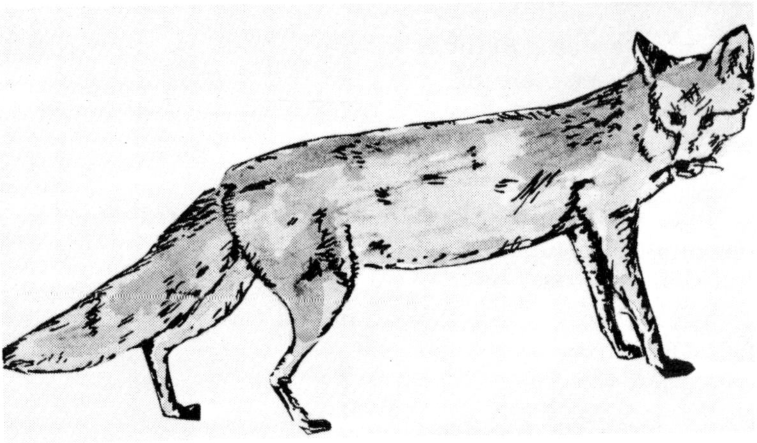

## SWIFT

"Wh, wha-, what was that?" the early-day working cowhand on the old Spur Ranch stammered, as an animal bolted in the brush and took off like a streak!

"Fox, I reckon," the poke surmised, but his cattle-punching buddy had this to say: "Not exactly, that's a 'swift.'"

"He shore as heck is," the first cowpoke laughed.

Both were right. There was a swift (kind of fox) out here in the early days.

* * *

Dr. W.C. Holden, in his book "Espuela Land and Cattle Co." wrote: "a sly, harmless little animal seldom seen was the swift. About the size of an overgrown Maltese cat, it belonged to the fox family.

"It developed speed for its defense and could outrun a horse."

Swifts were seen at times on several of the large ranches of the Caprock foothills.

* * *

There are many species of foxes, including gray fox, red fox, silver fox, blue fox, and even a white fox.

All species mentioned are natives of America. In addition, there is the yellowish-gray kit fox; and the swift fox, which lives on the dry plains of western America.

The gray fox is found on the plains of southwestern United States.

* * *

Foxes are relatives of dogs, wolves and jackals, but are smaller than those animals.

Their ears are larger and furrier, and shaped more like a triangle.

Their snouts are more pointed and their jaws are straighter. They have longer and bushier tails and longer hair.

* * *

Foxes gave settlers in this part of the country some anxious moments at the hen house. They are a sly and cunning operator, and tricky.

# # #

## EL LOBO
## (The Wolf)

The Llano Estacado is a great plateau of 60,000 square miles in eastern New Mexico and West Texas.

Roughly it extends from the Canadian River southward 400 miles, and between the Pecos River and Las Vegas, N.M., eastward to Palo Duro Canyon.

On it ranged many lobos, sometimes called "prairie wolves," and sometimes, "loafer wolf." They followed buffalo herds until cattle and sheep were brought out into that domain.

The lobos liked to follow sheep herds and pick out a straggler. They killed wholesale, not stopping until 25 to 50 head were slashed to death. They also stalked cattle herds.

* * *

W.W. Gunn, an oldtime cowboy at Spur, wrote: "Lobos came our way when I was on the Matadors. They were indeed cunning and killed a fresh beef a day. I trapped one of them.

"I have seen lobos from the Mashed O Ranch on the plains, to the Matador Ranch at the eastern foot of the Caprock. I was on the Matador Ranch in 1903. We always had to deal with the lobo, or 'loafer,' as we called him."

* * *

Cowpokes and ranchers and settlers on the Plains and below them, would spot a wolf or coyote den, mark it in their mind, and return to it when the work was up.

Sometimes after the male and/or female parents had been killed, the hunters would retrieve the pups.

John L. Wallace, Route 3, Post, wrote about a couple of tame lobos:

"I once has a friend, Willie Williams, who with his father was engaged in hunting and trapping lobos, back in the early days.

"They got $50 a pair for lobo ears, regardless of the age of the animal. They took a pair of young ones and made pets out of them. And used them to call up other lobos on moonlight nights. The captives would howl and call others up to within rifle range.

"Willie vowed they were never able to break lobos from killing a cow, if they got a chance. It always was necessary to chain the captive lobos to the wagon if there were any cattle around.

"My grandfather, J.I. Willborn, came to Borden County back in the 1880s. The first newspaper printed at Gail, July 31, 1891, gave an account of Willie Williams being in town with his pet lobos.

"This Willie Williams was a cook on the U Lazy S Ranch for years. One day, they were on a drive to Big Spring with a load of hides, and to buy supplies.

"A few miles out in the wilds, they came up on a little country school in session. It was a warm afternoon, and Willie's two lobos were trotting along in the shade of the wagon.

"A big, fighting Newfoundland dog had followed some kid to school. The dog came charging out, looking for trouble. He soon realized his mistake, and turned tail, running into the schoolhouse.

"The lobos ran to the schoolhouse door, but stopped. School was out, anyway! The kids saw the wolves and began jumping out of windows, and fleeing by the back door—not near big enough for the situation!"

# # #

## HARBINGERS OF TROUBLE

Horses of the two outriders were urged from a walk to a trot by the impatient knees of the cowpokes.

"Over there," Jake motioned, "circling."

High in the West Texas sky, the buzzards soared, down, then up, then turn, always circling.

"It may be the mother of that 'dogie' calf that showed up with the herd yesterday," Buck remarked.

"It could be anything or any one, 'dead, or dying!' Jake said, "we've gotta hurry!"

This was a common every-day occurrence out here in pioneer ranching days. The presence of buzzards circling in the sky, could mean trouble was lurking below them.

One good thing about this, help could be guided, even to remote pastures by the large circling birds in case of a prairie tragedy about to be discovered.

Fortunately this time, the carrion upon which the buzzards were feasting, was a dead rabbit.

And the range riders turned back for routine riding duties.

* * *

It is surprising that the bird we call a buzzard is not a buzzard at all. In this part of the country, his real name is "Turkey Vulture," but we call him "Turkey Buzzard."

He is one of nature's most effective scavengers.

The turkey buzzard is partial to the open prairies and barren hills and mountains of the South and Southwest.

He feeds along our highways. The turkey buzzard is not a shy bird, but he does move aside for oncoming traffic to pass. Then he returns to his cleanup work.

His flight is a thing of beauty. Few other birds have such mastery of flight. He needs only the slightest whim of air currents and crosswinds to keep aloft.

It is doubtful if the turkey buzzard ever attacks live animals or birds.

Most men recognize the importance of his work and do not shoot him.

# # #

## READY! AIM...

Buster Hicks, Lubbock consulting engineer and sportsman, drives 60,000 miles or so, a year. And on the seat of his automobile beside him, is—of all things, would you believe?—a slingshot!

When he goes into all kinds of terrain, to inspect an engineering project, Hicks things nothing of stopping to kill a rattlesnake, or knock off a rabbit—with one shot per target.

He's good all right. After all, he has been carrying the slingshot for 30 years.

"I guess every small boy in the world, sometime or other, has made, and shot a slingshot. I just kept mine," Hicks laughed.

"When I was a boy down on the farm-ranch in Comanche County, it was a way of life. Now, it's my hobby," the Lubbockite explained.

His slingshot is the regular forked limb for a stock, two strips of rubber, and a piece of leather-variety, but it is a dandy—sort of a "sophisticated" or "engineered" version.

The well-worn stock or fork, is made of cedar. The two rubbers are cut from electric linemen's gloves. The rock or pellet holder is made of soft leather. The rubbers are tied to the stock with fishing line, and the strings from the rubbers to the rock holder are nylon.

"I use ball bearings, rounded lead sinkers from fishing lines, or smooth round rocks, if I can get them—for ammunition," Hicks said.

"I have a lot of fun with this thing," he said, "I take it along to the golf course in my hip pocket and nobody notices it." the engineer shooter said.

"Then, when a friend goes into his backswing for a drive, I sometimes pot shot the ball off the tee. You ought to see their faces when they swing down and discover the ball gone from the tee.

"I was playing in a foursome with Dizzy Dean and others in the Donny Anderson Celebrity Tournament (1972) and happened to see a ground squirrel about 100 feet down the fairway. I drew back with my 'shooter' and fired, hitting the ground squirrel the first shot and it plopped over dead.

"Old Diz let out a whoop—'that's the shot of the day, and it ain't no golf shot, either!'" Hicks quoted Dean as exclaiming.

"Course a fellow gets lucky and knocks off a difficult target and from then on, he gets credit for being a crack shot," Hicks confessed.

But his friends say, Hicks is "lucky," then, each time he 'fires' the slingshot.

* * *

Ground squirrel is the name for a burrowing member of the squirrel family, who digs his home as a hole in the ground.

The common ground squirrel of the U.S. has 13 striped lines along its back. There are 97 distinct types of grond squirrels.

Ground squirrels were seen frisking about this part of the country in pioneer days, but their numbers have diminished greatly. Nowadays, they are a rarity.

# # #

## MRS. WREN'S NEW HOME

Here's a story for the birds:

Mrs. A. Wren flipped saucily in a three-toed walk along a backyard phone line, where she had been "chirping" things over with the girls—Mrs. S.B. (Song Bird) Mockingbird; Mrs. Justa Sparrow, and Miss Ima Lady Blue Jay (not yet mated).

This happened in Lockney, Tex., about the year 1966.

"I just must have a new 'house,'' Mrs. Wren chirped. "Only last night, I told my Abie (Mr. Wren) so.! Ours is such a drafty old place in the apple tree.

"Peep, peep," she cheeped: (meaning "only the other day, I found this darling little ready-built place over by McDonald Construction Co.")

"Cheep, cheep, cheep-cheep, and theep!" clattered Mrs. Mockingbird and Miss Jay: (meaning "much too small for us—such little bedrooms, don't you know!)

"Chirp, chirp!" from Mrs. Sparrow (meaning "too ritzy for my brood. We'll have to get along with last year's string-thatched nest.")

"Ka-peep!" Mrs. Wren again: (Meaning "you're both just jealous. I'm flying over right now to look at one.")

* * *

Which is all to point up how a Lockney furniture dealer-insuranceman, from his hobby shop, has established a production line of birdhouses that extends to Lubbock, Amarillo, Dallas, and other cities.

"Art's Originals," they are called. "We try to make each one different," explained Art Baker, Jr., architect-designer, who contracted McDonald Construction Co.,—to help produce the bird houses.

The little houses are in three styles—"single family," "larger family (two and three-bedrooms)," and one with a little red barn.

They are made of cedar wood and each is stained and antiqued.

Each has a small rock doorstep, red and green shutters.

Windows are neatly screened and glassed.

On each porch is some object, like a broom, churn, wood pile, or water bucket—for a "home-like" atmosphere.

# # #

## OLD "BUCK TOOTH"

The story had been bandied around the cow camps from Arizona to the Texas Panhandle, before it came to the cattle country by the Caprock.

It was a tale about an old rancher and "Buck Tooth," a sidewinder (a kind of small rattlesnake but a vicious one) and how they picked up a small fortune with a little betting stunt.

Old "Buck Tooth" was kept in a glass jar with a buckskin lid that had air holes in it.

The rancher-owner would come into a place, deposit "Buck" on a table or counter, and then clean up, betting all comers (at a dollar a try) they could not put the palm of their hand on the side of the jar and let old "Buck Tooth" strike—without withdrawing the hand.

But the rules specified the bettor had to look at the snake, right into his lidless, unblinking reptile eyes, dark and sinister.

Many tried it. All failed. None could keep the hand from jerking away as the rattler struck with its evil-looking triangular head.

Some tried several times, even getting "fortified" with a little "red-eye," before betting again. The result always was the same—the hand would jerk away no matter how hard the bettor tried to keep from doing it.

One cowpoke tried holding his palm down with his other hand. It didn't work. He flinched when "Buck" coiled and struck—both the palm and the supporting arm came away from the jar.

They said it was hardest when the old rattler didn't strike immediately, that beads of "cold sweat" broke out on the forehead of the man with his palm laid down on that glass jar, watching that snake...waiting... waiting...waiting!

There wasn't much knowledge of physiology in those days, so they just laid this to some kind of "reflex action."

Anyway, they tried it with a diamondback down in the Spur country. And we saw it. The results were the same as in the handed-down story.

Several tried that West Texas rattler. Some pretty nervy guys did. But when the rattlesnake let go a strike at their palms—no one could keep a hand on the glass (they used a glass case).

And what ever happened to "Old Buck"? Well, they say one night in Dodge City, a wild cowpuncher lost three bucks straight, betting he could beat the game. He lost.

So he got so mad, he hauled out his six-shooter and with his first shot broke the glass jar, and with the second, sheared off the head of the rattler.

The owner screamed "bloody murder" at losing his money-making game and had to be restrained from bracing the cowpuncher that killed "Old Buck."

Just a story? Perhaps, yet we saw this thing tried with a diamondback rattler. No one could keep his palm down when the cruel-eyed rattler struck.

Can you?

# # #

## CITY BIRDS

"Daddy, look at all the 'black birds' on the courthouse lawn," a small West Texas boy said to his father as they drove by on the way to the boy's school on a day in fall.

"Those are starlings, son, although they are black," the father replied.

As he drove on to the school, the father recalled when starlings by the hundreds invaded cities in West Texas several years back.

So thick were these "visiting populations" of birds, special shooting events were held in at least two cities—Midland and Lubbock.

The starling "hunt" at Lubbock was held on the campus of Texas Tech. The campus was blocked off from automobile and pedestrian traffic, at the time of the organized shoot. Shotguns were used.

At Midland, the flocks of birds chose the courthouse at which to converge. Shooting parties were authorized and supervised.

* * *

The common starling is the kind we have in West Texas and we only see him when the flocks are passing through. He is a black songbird and has a long, sharp bill.

Starlings are very helpful to farmers by destroying insects, but in fruit season, they can become pests, eating many berries, cherries and even apples and pears.

Starlings, roosting in city shade trees and in parks and about buildings, in fall and winter, can be a great nuisance.

* * *

About 80 common starlings were set free in Central Park, New York City, in the late 1800s. From this small beginning, they have spread as far south as Texas and Florida; and as far west as Utah.

* * *

Actually, the little West Texas boys's calling starlings "black birds," wasn't far from correct, although he merely was referring to color. The starling, indeed, is of the black bird and grackle family.

All of these—starling, grackle, black bird—have the same tendency to "travel," roost, fly, and stay in flocks or bunches.

# # #

## RABBIT DRIVE

Rabbits, both cottontails and jackrabbits, were on the Plains and in the Caprock country when the first settlers arrived.

They were not nearly so plentiful then, as later, when the wolves and coyotes were thinned out.

Rabbits were no problem in the time of Indian occupation of this area, or during the buffalo hunter's era.

But as the settler advanced, he killed off the predators (wolves and coyotes) that had helped keep the balance of nature.

Rabbits then flourished and multiplied, as rabbits can do. They became a menace to gardens, young fruit trees and even crops.

There arose an activity in West Texas called the "rabbit drive"—designed to get rid of the animals when they became too numerous.

This idea was gained from the Indians. One of the redskin's favorite methods was to form circular or V-shaped lines over a wide area where there were many rabbits. Then, the walking hunters gradually would come toward each other, forcing the rabbits into a small space, where the Indians killed them with clubs.

The white man refined this method. He used shotguns to kill the rabbits. Unlike the Indian, who used rabbits both for food and fur, the white man had no use for a dead rabbit.

Down around Spur, a rabbit drive was quite an event. It would be advertised in the newspaper. Usually a meeting point would be a rural schoolhouse.

The "drive" would be highly-organized and safety precautions were taken. The men would form a line a section or half-section long. At each end of the line would be stationed a captain. These leaders were to give the "advance" or "hold" signals.

A group of shooters would be sent on ahead of the line of walkers by a round-about route. These stationary shooters held their position as the line walkers advanced toward them, driving the rabbits.

The advance guard or hold men were allowed to shoot until the line of walkers came within range.

Each man carried a shotgun. Sometimes, a farmer or rancher would even furnish the ammunition to get rid of pesky rabbits.

In the line, men were set 60 to 80 yards apart. You shot as you advanced and saw targets. No dogs were allowed. It was a cardinal rule not to go forward in advance of the line.

Even so, there were some victims who had to have pellets out, now and then. Usually, there was a doctor or two on the drive.

There was little or no danger of a fatality, if the spacing alignment was kept.

At noon, the women folk would have a big meal ready at the schoolhouse or some designated spot.

This always was a great attraction. The food wouldn't be "rabbit stew," either, but barbecued beef, tasting as only that kind of meat can, cooked and smoked on a mesquite fire.

# # #

## THE LOVELY "MEDDER LARK"

William Laurence (Larry) Chittenden was no "dude." An Easterner, he caught the flavor of the West in poetry—like most Easterners didn't.

Chittenden, from Montclair, N.J., worked as a reporter in New York City. He migrated to Texas in 1883, with an uncle, and established the Chittenden Ranch, near Anson.

He began writing western poetry and it gained national recognition while he was working 13 years as a cowhand.

His best work was "Ranch Verses." It went through 16 editions.

Many say his best poem is "Cowboy's Christmas Ball," that starts something like this:

*"Way out in western Texas,*
*Where the Clearfork's waters flow.*
*Where the cattle are a-browzing'*
*And the Spanish ponies grow;*
*Where the northers 'come a whistlin'*
*From beyond the Neutral Strip;*
*And the prairie dogs are sneezin'*
*As if they had 'the grip';*
*Where the coyotes come a-howlin'*
*Round the ranches after dark.*
*And the mockingbirds are singin'*
*To the lovely 'medder lark' "* . . . .

Right there, Larry Chittenden mentioned one of the favorites of West Texans—the meadowlark, or field lark, he is sometimes called.

* * *

The meadowlark likes to perch on a fencepost and make the announcement—"tee-you, tee-yair, tee-you, tee-air!"

A loud bubbling refrain bursts from his throat to tell country folk, "Spring here, Spring is here!"

This "old friend, the field lark" is a gay minstrel.

About the size of a robin, the meadowlark is a brown bird with darker streaks on the back, a black crescent is emblazoned across the yellow breast.

With a vibrant, rollicking song, the meadowlark inhabits the prairies, fields and upland meadows of the West.

Sometimes, if enough material is at hand, the grass-woven nest in a depression of an open field has a covered walkway, with a secret entrance.

Once, the 'lark sang out here in fairly large numbers. But now, he is seldom seen, and if so, it is rare—far from the cities' rattle.

# # #

## ANIMAL ENGINEER

"They had a dam up this creek in Blanco Canyon about where the Bridwell Ranch straddles the Caprock edge," the old surveyor said.

"Beavers," he added as a one-word explanation.

The surveyor was the late George M. Williams of Spur, who worked mostly for the Swenson Ranch. As a boy, this writer walked as his chainman.

We were running a little review survey out west of Spur, on what was then called Catfish Creek (now White River).

At the lunch break, cooling off in the shade of a mulberry tree, we asked Williams for some more talk about those beavers. We had never seen one, nor heard of one being around in West Texas, and for sure there weren't any building dams at that time (1920s).

"They were great little engineers," Williams commented, "and hard workers, too. Water poured over their dam for years and backed up a good-sized lake."

"It was about 2 miles upstream from Silver Falls, once a recreation area and now a small fall over a concrete dam beside U.S. Hwy. 82," the old surveyor added.

* * *

Later, we learned that as late as 1880, beavers were found widely scattered along running streams, where cottonwoods grew from the Caprock to the 100th meridian.

The Bridwell Ranch was carved off the corner of the old Two Buckle Ranch, one of the early spreads of the Blanco Canyon and Caprock foothills area. The Bridwell land was first owned by Sidney Webb, who got it from the C.B. Livestock Co.

This is the Blanco Canyon area, east of Crosbyton, cradle of Plains civilization. Here is the hulking skeleton of the old Hank Smith Rock House, built in 1878, and also the site of the rock ranch house of the Two Buckle, erected in 1885 by the Kentucky Cattle Co.

This was the first ranch house of the South Plains area and now is the ranch house of Dr. O.W. English's Bar X, at the Silver Falls roadside park.

* * *

"Those beavers were trying to tell us something," the old surveyor concluded.

Indeed they were. First, Webb built a dam in 1923 on White River and backed up a lake that was the center of a recreation mecca for years. Next came the White River Dam, 12 minutes south on White River, which furnishes water to four area towns—Spur, Crosbyton, Post and Ralls.

East of here, around the old Spur Ranch headquarters, and the Guy Goen Ranch, man-made lakes are being hewn and precious water is backing up toward the Caprock like the beaver lakes used to do.

But the busy old beaver is gone now, and has been for many days. Few modern-day residents know he ever was here.

However, he left an idea—that of catching water with dams that is proving of great benefit to man today.

# # #

## THIEF! THIEF!

The neophyte rustler wasn't too sure of himself in the first place.

He had rustled some choice steers from a big ranch when he thought no one could be watching.

Now he was miles away in a secluded thicket and was taking it easy to rest his horse. The stolen steers were munching away on a good stand of grass near a spring of water.

But he kept seeing flashes of blue and white as a flock of birds gave out, now and then, with a raucous call that sounded like "thief! thief!"

Now Mr. Rustler wasn't a native out here and he knew nothing about the Blue Jay.

The noisy birds, full of sass and vigor, kept it up, louder and louder: "thief! thief!"

The rustler was getting nervous by the minute. Although he thought, "I'm silly, no person or animal or bird, saw me rustle them steers. But why does that pesky bird keep chanting—"thief! thief!"

Finally, he could stand it no longer and decided to head his small ill-gotten herd back to the owner's ranch, to report he had found these strays and was looking for the owner of the brand.

Just a humorous little story? Sure, but knowing the Blue Jay makes the humor of the tale appreciated.

* * *

The Blue Jay is a handsome fellow with his bright blue wings and tail, black collor and white underparts. And he is topped with a crest.

He bluffs and he blusters, all right. Sometimes it sounds as if he is calling "Jay, Jay, Jay," as if to call attention to his noisy presence. He scolds and he screams as he moves from grove to grove, and from thicket to thicket.

This fellow delights in terrorizing smaller birds by mimicking the cry of the sharp-skinned hawk.

Generally, Blue Jays move south in the winter. Yet many are residents the year around in northern climates. As some move out to the south, others move in, with arrogance and bad manners. Nevertheless, the Blue Jay is a beautiful and amusing bird.

# # #

## BURRO SENSE

"I've got you, you long-eared jackass, you, I've got you!" the excited cowboy yelled.

He quickly mounted and that's where the fun, or rather the action, began.

One of the most interesting and humorous stories of the Caprock foothills was the one that related how a cowpuncher with an early-day cow outfit was by himself, when he lost a horse that stepped into a prairie dog hole and had to be shot.

This time, the poke had no remuda with him and he was miles and miles from the ranch headquarters, which were down around Anson.

The "up-the-creek" cowpuncher, as luck would have it, jumped on a wild donkey that was drinking at a remote spring of water up against the Caprock.

After much trial and tribulation, the cowboy who had been "a-foot," rode the donkey (or burro) back to the ranch headquarters, several days and many bruises and balks later.

Needless to say, the puncher got quite a ribbing from the boys at the bunkhouse.

"Old Buck done captured himself a wild mustang up there in Caprock Country," teased a bunk mate.

Buck, sore in muscle, body and pride, didn't have much to say.

* * *

Donkey or burro is the name of several animals, that are relatives of horses. But they are smaller and sturdier.

The wild ass of Abyssinia and northern Africa is the ancestor of the common domestic donkey.

The smaller donkey or burro is a fine pack animal because he is sure-footed and can stand all kinds of weather.

* * *

Fred Arrington of Dickens in his book "A History of Dickens County," tells about a deskmate he had one time at Prairie Chapel School who rode a donkey to school:

"The teacher asked him to write a story about donkeys or burros. And Aaron did it up great...

"He wrote 'Horse sense just don't stack up to burro sense. The burro may be the world's country cousin, but he is plenty smart...For one thing, the horse is ignorant enough to submit completely to man... A cowboy can keep pushing a horse until it drops dead. Did you ever hear of a burro doing that?'

" 'A burro has the patience of Job...You can't push him into a bog or quicksand...I have seen a horse tangled in barbed wire. He lunges and tries to get out. A burro, if he is careless enough to get tangled, waits quietly for someone to help him.'

" 'The little burro won't gorge himself on food or drink too much water; a horse will.'

" 'The Bible tells where the Master rode a burro into Jerusalem. The burro has been man's friend from the beginning of time...' "

# # #

## BUTCHER BIRD

Boys growing up in the ranchlands of the Caprock foothills soon become acquainted with the "butcher bird." That's his common name; actually he's a shrike.

He gets his name from his unique habit of impaling his catch—grasshoppers, crickets, field mice, small lizards, etc.—on thorns or wire barbs, much as a butcher hangs meat.

There are reasons for his doing this, encyclopedias and nature studies tell us:

Like all birds of the passerine group, shrikes have feet adapted to walking, hopping, and perching. But its claws are too weak to solve the problem of tearing its food prey into pieces small enough to be eaten.

So Mr. "Butcher Bird" has developed the trick of wedging his food supply in the crotch or fork of a tree branch, or, more often, impaling it on some sharp object so that he can tear his food into pieces with his strong beak, or, if preferred, his bill.

Shrikes are wholly carnivorous. They have been known to impale a fish, although frogs, field mice, lizards and insects are more to his liking.

The "butcher bird" often hangs up his catch when prey is plentiful, and returns to it, when prey is scarce.

Shrikes hereabouts are a medium-sized bird, up to 10 1/4 inches long. They are black over gray, or off-white in coloring. The well-developed beak is slightly hooked and very strong. The bird hunts like a hawk but kills with his beak instead of claws.

He's a keen hunter, and swift, making fierce and frequent sallies after prey from his favorite perch. Insects and smaller prey sometimes are devoured on the spot of catch. Larger prey is impaled.

The "butcher bird" is one of five so-called characters. Sometimes, he is called "highwayman of the air."

The other characters are: the crow, the great horned owl, Cooper's hawk, and the sharp-shinned hawk.

Shrikes have few likeable traits. They are rascals, sometimes nicknamed "Bluebeards." Although the family is classified with songbirds and although he tries hard to warble, the shrike's call is harsh.

Wherever he lives, he occupies a conspicuous place among feathered folk of the countryside, say the nature study books.

Farmers and ranchers tolerate him, though, because he destroys insects and harmful rodents.

# # #

## WHOA, BECK!

Mules were valuable domestic animals and faithful beasts of burden.

They were used by early-day ranchers in West Texas to pull "chuck" and "hoodlum" wagons. The mule wouldn't scare easily and was more dependable than a horse.

The mule was to the farmer what tractors are today.

They were great workers in the heat. And were the best pack animals you could get.

The mule had his day, all right, but now his kind are all but gone.

* * *

Appropriately, the lowly mule has been honored.

In July, 1965, a handsome black Fiberglass statue of a mule was unveiled in Muleshoe.

The "mule memorial" is to honor, at long last, the faithful beast of burden, who broke sod on the rich agricultural Plains and other areas of West Texas.

The memorial, which includes an official Texas historical marker along with the statue, is on busy U.S. Hwy. 84 and is turning out to be a real tourist attraction. It also is pleasing to the homefolks.

The historical marker points out that "the mule, along with buffalo, hound and Longhorn—made Texas history..."

A dedication speaker said: "There must have been many a day when the early settler said to himself, 'I'll hang on for just one more crop season. If my mule can make it, so can I.' "

Gil Lamb, Muleshoe radio personality and operator of Station KMUL, became president of the National Mule Memorial Association. He pushed the mule memorial and marker in his popular program "Mule Train."

Lamb said: "Mules plowed the first sod for pioneer man...Ground the first meal for the pioneer woman...Mules built the first railroads westward...Mules pulled the covered wagons west...Hauled the first freight...Delivered the mail...Mules built the first highway roadbeds...Mules, 5,000 strong, served in World War I."

* * *

Now comes Arch Lamb, Lubbock County Commissioner, in 1972 and breeds a sorrel mule, the first mule colt born around Lubbock in years.

Arch asked Dr. Fred Harbaugh, who was Texas Tech's veterinarian more than 43 years, if a sorrel mule was rare? "Arch," Doc Harbaugh said with a smile, "any mule around here is rare today!"

W.C. Homsley, herdsman at Tech 14 years, and Raleigh Middleton, another oldtimer, reminisced about mules, particularly those at the college (now university).

They recalled that mules were worked at Tech in its early days. And as Homsley remembered "they made fine draft animals. Even Mr. Grimsley, the custodian, drove a team of mules to a wagon for campus pickup."

Many today, who came from the farms, may recall following a mule many a mile down a row, hitched to a one-row cultivator, or even a "Georgia Stock" plow. And that most mules had names, ranging from "Beck" and "Jude," "Lucy" and "Blue," to "Scissor-Mouth," an old mule that served at Texas Tech.

Remember?

# # #

## CRANES OF THE SANDHILLS

Early settlers, moving West, referred to the Plains of West Texas as a "sea of grass." They were right!

Some claimed the area was fit for neither man or beast. They were wrong!

Landscape on the Muleshoe Wildlife Refuge is very much like that which greeted early settlers of this area. Less than 3 per cent of the 5,809-acre refuge has ever been cultivated.

Contrary to belief of some settlers that the land here is unfit for man or beast, the refuge is inhabited by thousands of migratory and nesting birds, and annually, thousands of persons come to see them.

The Muleshoe Refuge probably is best known for having the largest concentration of sandhill cranes in the U.S.

The big birds arrive on the refuge during the third week in September. They remain for about six months.

They come from their Arctic breeding grounds, as far away as Siberia.

The number of these birds on the refuge reached 95,000 in January 1971.

There are three large lakes on the refuge.

The Muleshoe Refuge was established in 1935 and is the oldest national wildlife refuge in Texas. More than 7,000 persons visited the refuge in 1971, utilizing camp facilities for bird watching and picnic outings.

Many of the cranes used to show up all over the Plains and down in the Caprock foothills. And still do.

When the short hunting season for sandhill cranes comes around each year, the Muleshoe Jaycees stage their annual Sandhill Crane Hunt and take hunters out to the spots just outside the refuge, where some of the big birds swoop down on tanks and lakes.

But most of the birds are "smart" and fly around only in the refuge, safe from hunters' blasts.

In earlier days, before the refuge was established, the cranes were seen in many parts of the Plains and Caprock foothills country.

One of the places they always seem to frequent are Tahoka Lakes, although numbers now are much smaller than in earlier days.

# # #

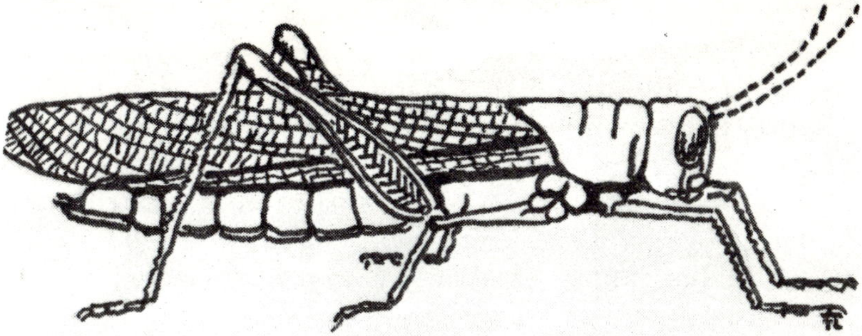

## THE PLAGUES OF WEST TEXAS

Anson Cox, one of the earliest settlers of the Southern High Plains, once told of a severe grasshopper plague in 1893.

Cox first discovered the ground was perforated with little holes in late 1892. He said the next spring, as he was plowing, "millions of little things about the size of fleas at that time, hopped up and down."

Upon examination, they proved to be young grasshoppers. They soon grew large enough to be noticed everywhere, and began eating "everything green in sight."

According to Cox, one farmer near Estacado, first settlement on the Plains, had a beautiful field of wheat. The hoppers soon devoured all grass and turned to fields. The farmer vowed they would never eat his wheat. He had a large family and he armed each member with a brush-broom to kill the grasshoppers as they passed a furrow he had plowed around the field of wheat.

Then, he hitched his team to a large log and drove the horses at a trot, crushing the hoppers as they bunched in the furrow. But in vain. After an hour or two, he gave up.

The 'hoppers came by the millions and moved on, always to the east. Cox wrote he thought he could save his garden by filling a ditch around it with water, but the 'hopper horde marched in, filled the ditch and others went right over the drowned ones.

Cox reported there was devastation everywhere. Anything green in sight soon was gone. It looked, he said, "bare and as if no vegetation had ever been there."

* * *

About 80 years later in 1972, like phantoms from out of the past, a massive grasshopper invasion was reported at Midland.

City-county health department crews dispensed insect spray in almost a steady stream, "killing the pests any way they could."

The insects, thousands of them, some as long as 3 inches, invaded the Permian Basin city at night, interrupting an Amarillo-Midland Texas League baseball game. And the second game of a scheduled doubleheader, was called in the first inning.

Women screamed, players swung their bats, fans clawed at the grasshoppers that soon covered the playing field, the stands, and the mercury vapor lights.

The invasion was not confined to the ball park. The sticky hoppers were found throughout the city—flying, crawling, hopping!

A Dept. of Agriculture agronomist said the hordes of hoppers at Midland, probably came looking for sorghum and wheat grain since dry weather had left barren fields.

The hoppers advanced closer and closer to the city, searching for grain and growing vegetation.

Once inside the city, the grasshoppers went after grass and shrubs and gardens.

# # #

## SYMBOL OF PEACE

Oldtimers said there always were "a world of doves" along a country road at the foot of the Caprock that leads to the Spur Headquarters Ranch.

This has been true in our lifetime for we have never been along that road that skirts the old Espuela townsite now quilt-patterned into farms, without seeing doves, in season.

At least two creeks in Texas are known as Dove Creek. One rises in Dickens County and flows southeast 18 miles to join the Salt Fork of the Brazos in northern Stonewall County. The other rises in Schleicher County and flows 37 miles across Irion County and into Tom Green County.

There is a Dove Mountain in east-central Brewster County.

And there was a Dove Creek south of San Angelo in pioneer days that figured in an Indian fight. It occurred Jan. 8, 1865, 11 miles south of San Angelo and was called "The Battle of Dove Creek."

Some 1,400 Kickapoo Indians were overtaken by Capt. Henry Fossett and 370 State Border guards and a battle ensued. The Indians got the best of it in a desperate fight. The white men lost 36 men killed and 60 wounded. The Indians, under Chief No-Ke-Wat, lost 11 killed and 31 wounded.

The dove is an interesting bird. The Bible tells us the dove has been around since the days of Noah and the flood. It is related that Noah sent out doves from the Ark to find out if waters had gone down.

* * *

Some people think Mourning Doves have nothing to do all day but "bill and coo," but on the contrary, they are busy birds. They raise not one, but two or three families in a single season.

All through spring and summer, the tender, sorrowful mating song of the Mourning Dove, is heard—"coo, coo, coo!" From these sad notes, the Mourning Dove has been named. Yet this dove is not unhappy.

MDs are poor home builders. Their nest is a small platform of twigs, crudely laced together. They are fortunate if two white eggs do not fall through a hole in the unlined structure.

Father Dove sits on the nest in daytime. And Mother Dove sits there at night.

The Mourning Dove has an olive-brown body, bluish-gray wings and plumage. They are about 10 inches long and have a long-pointed tail with white-tipped side feathers, distinguishing them from other members of the pigeon and dove families.

Mourning Doves gather in large flocks to make sprightly flights on whistling wings in preparation for an annual flight south for the winter. Many times, they perch on telephone lines and swing in the breeze.

The hunter has not decreased their sizeable numbers to a great extent.

# # #

## CLOSE "CALL"

Cecil Fox, retired game warden at Spur, is a bobcat-hunting man, but one time, a "wild kitty" got too close for comfort.

Cecil is an expert at making and using mechanical "callers"—devices that give out the cry of a wounded rabbit.

Coyotes and bobcats, even hawks, come running, or flying, in anticipation of an easy meal when they hear that sound.

Fox and rancher Horace Wood were concealed in brush so as to not be seen by the predators.

Fox was stretched out full length on his stomach and blowing lustily on the "calling" device. Of course, his rifle was handy and Wood was watching another direction.

The "call" was sounded several times. And then came the lull, and silence. Fox was watching straight ahead and

Wood to the sides of their improvised blind, located in rugged ranch country at the Caprock edge.

Sure enough, a big bobcat had heard the call of the wounded rabbit, and was curiously investigating.

It just so happened, Brother Bobcat approached from the rear, and seeing Fox prone, he couldn't quite figure out what was what. So, the curious 'cat went around Fox and approached right at his face.

"Eyoweeeeeeeeee, scat! You overgrown house cat!" Fox exclaimed, scrambling to get on his feet and to bring his rifle into position.

The bobcat, about as surprised as Fox, jumped about 2 feet high, let out a squall, and took off.

Fox, a crack shot, downed the bobcat on the run up a nearby slope.

"Ha, ha!" laughed rancher Wood, "why didn't you pet him, Cecil, he was close enough. All cats like to have their heads scratched!"

"Oh yeah," Fox countered, "I didn't notice you offering him a saucer of milk, Mr. Wise Guy!"

"Hey, look at them claws," Fox said, as he walked over to get his trophy, "sure glad I didn't scratch his head, he might have wanted to return the favor."

* * *

In more recent years, mechanical "callers" have been developed that work very well to bring up prairie predators.

Hunters with patience and the right conditions, find good targets in their sights when the distressed rabbit call is given.

It is the same principle employed by duck hunters with their whistle-like callers; and moose hunters with their deep-throated horns.

# # #

## DUCK CREEK

Duck Creek is the name of at least nine Texas streams.

The best known one in West Texas is the Duck Creek that rises at the breaks of the Llano Estacado in northwestern Dickens County, and flows southeast 40 miles into the Salt Fork of the Brazos River in central Kent County.

This Duck Creek got its name from the many wild ducks that came to the water holes on it from the north in the fall of the year.

Much history has coursed along Duck Creek, so to speak. Spring-fed, it flows between the headquarters of the old Spur Ranch and Soldier's Mound, where Mackenzie's Raiders had a supply camp in the 1870s.

Much of the creek's course was in a pasture of the vast Spur Ranch spread.

The Spur region was great buffalo country. In 1876, many buffalo hunters camped on Duck Creek.

But Duck Creek's history goes farther back than the buffalo hunting era; for many years, it had been a place of rendezvous for the Comancheros and Plains Indians.

In 1871, evidence of the Comanchero occupancy on Duck Creek was found by an expedition of Gen. R.S. Mackenzie. While riding along Duck Creek, Capt. R.G. Carter reported finding trading stations—"curiously-built caves in high banks and bluffs." These were later identified as dugouts, and the same type of dugout was used as living quarters by the first cattlemen in the Caprock foothills.

It is noteworthy that in the year 1972, modern projects on Duck Creek by the Soil Conservation Service to prevent water erosion and flooding, including several dams, were dedicated.

* * *

Interesting about ducks, either Teal or Mallard, is the practice of the female of carefully covering her nest with down or leaves when leaving for a short swim, while nesting.

Ducklings only a few hours old can swim and run swiftly. At six weeks, they can fly.

Many of the large flocks of wild ducks formerly seen in this area have disappeared. Yet, despite hunting and destruction of nesting sites, they still are quite prevalent. Game reserves and refuges shelter many.

# # #

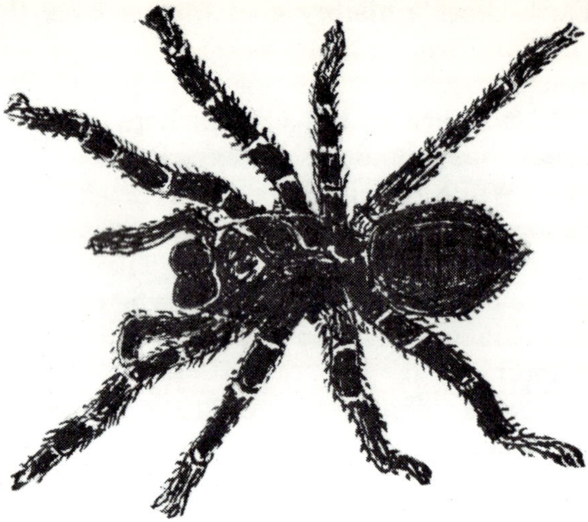

## PECOS BILL'S PETS

Texas means the cowboy and his customs to a great majority of tourists and visitors.

The Texas cowboy bears the same relation to folkways and folklore that the Indian does to Oklahoma.

That feeling of earth and wide sky while tending the herds, the "yipiee" of the rodeo, the acrid smell and lusty sounds of the roundup, and the hardships, accepted half-humorously, half-resignedly, have been factors in the development of cowpuncher lore, which has survived even the passing of the big ranches.

Why, it's only natural that a mythical super-cowboy would have evolved. Yep, "Pecos Bill" was his name—the granddaddy of all cowhands!

He experienced such tremendous adventures, he even altered topography of the state.

Any cowman can tell you "Bill" was born in Texas. He

dropped out of a wagon one time, when his folks were moving farther west.

Coyotes raised him. He became so tough that rattlesnakes hid in the cactus for fear he would bite them.

He used mountain lions for saddle horses, a live rattler for a lariat. Centipedes, scorpions, vinegaroons and tarantulas were his pets.

On a bet, "Pecos Bill" mounted a cyclone and traveled far and wide. From his little ride, mountains were leveled and forests uprooted. The almost treeless Panhandle and Plains emerged. That's why it's so flat out here.

Yessir, "Pecos Bill" was quite a cowhand!

* * *

## TARANTULA

Largest of the spiders. Any of the large, hairy spiders found in the Southwest are called tarantulas. They do not catch their prey in a web, but capture enemies by grasping them, and then eat them. Tarantulas live in little wells in the ground, lined and covered with a silky web. His bite is painful but is not considered dangerous. No one has ever known a person to receive a fatal bite.

* * *

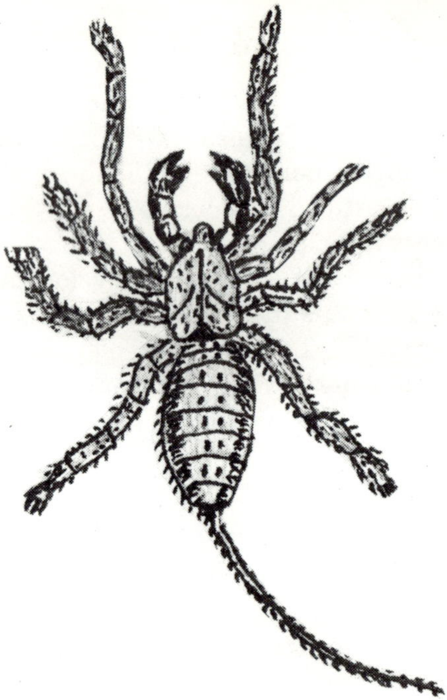

## VINEGAROON

So called from its odor, when disturbed. It is a large scorpion found in the Southwest and in Mexico.

SCORPION—Has a poisonous sting in its tail. It is common in the Southwestern part of the U.S. It stings its prey and sucks the body juices. There are 20 different kinds of scorpions in the U.S. The scorpion has six to 12 eyes, depending on what kind he is. The stinger is in the tail, curved at the end of the tail. The constellation "Scorpius" forms a scorpion with stars. The bright star "Antares" is near the center.

# # #

## BOB WHITE

There is no thrill in modern urban living to compare with being in the countryside on a fine morning or late afternoon in Indian Summer, hearing the introductory greeting from an unseen feathered friend—"Bob White, Bob White, Bob White!"

Yes indeed, a cheerful friend of the farmer and rancher is Bobwhite, the quail.

He was here once in teeming numbers, and still is around in remote areas, where red-capped, lace-booted, shotgun-armed sportsmen will be hunting him in season.

But this is a different story, a closer look at this interesting fellow we know as "Bobwhite."

He is a plump bird, slightly larger than a meadowlark (field lark). He wears a coat of brown and buff. A white line slants across the face and over the eyes. These, and a pure white throat, are his markings.

Bobwhite is a kind and considerate companion. He shares sitting duties with his smaller mate, where she has carefully arranged 10 to 20 eggs in the nest, always with the small ends down.

Each morning for 23 days, he sings to her from a nearby thicket or tree limb.

Baby bobwhites are quaint and lively from the moment they hatch. Parent Bob can give a warning call of alarm in case of danger that sends the little ones scurrying to hide beneath their mother's "skirt."

As soon as the young birds are feathered out and have grown some, the family roosts in a tight circle on the ground, tails to the inside and heads facing out. They are constantly on the alert for coyote, fox, hawk, or other predators.

They hug the earth closely and seem to become invisible by one of God's and Nature's miracles of camouflage. When flushed they explode into flight like a bombshell! Their flight is spirited and strong.

The bobwhites have their own private walkways through underbrush. The babies follow the parents in single file on distinctive, well-worn paths from field to brush and back.

When the baby quail are fully grown, the family joins other families to form a bevy or covey. These groups stay together until spring mating or pairing season rolls around.

We have heard farmers and ranchers declare that bobwhites sometimes seem to call out and sing of warm days and more rain—"More wet, more wet, more wet!"

# # #

## POLECAT

Of all wild creatures, the cowman gave the widest berth to skunks—or "polecats" as he called them.

No varmint that roams the woods or ranges commands as much respect as the skunk with his built-in defense mechanism.

Both men and other animals, as well, give him plenty of room.

For one thing, according to the late George Lightfoot of O'Donnell, "polecats could stampede a herd. I saw one do it. That old black and white 'kitty' came ambling up with his 'flag' raised straight up. An old steer snorted and that did it. The herd lit out."

* * *

"I don't care what you say about his being helpful by catching rodents and insects, I don't like one bit one of them fellers that walks with his back arched and steps along like a person whose boots are too tight!" declared an old cowhand.

Skunks sometimes were rabid and their bite could produce the dread "hydrophobia," as the pioneers called rabies.

* * *

Vern Sanford in his column "Outdoors In Texas," came up with some interesting research on skunks. He found one resident of the great outdoors that didn't fear Mr. Skunk. It is the great horned owl.

"This old bird now and then enjoys dining on a tasty skunk. Obviously, the owl has no sense of smell and could care less about the odor," Sanford wrote.

But other birds, animals, or reptiles, as well as man, steer clear of the skunk, whose odor can linger for weeks.

Sanford said there is no sure-fire method of removing the tell-tale skunk odor. The liquid spray of skunks contain an ingredient known as nubutyl mercaptan. And contains sulphur.

One of the best antidotes is said to be washing in liberal amounts of tomato juice.

As for doused clothing, about the only thing to do is either burn or bury. Even then, the odor seems to linger on.

The skunk is a descendent of the weasel family. And they say, one makes an excellent pet when the scent sacs are removed.

The pioneer used the name as a term for bad character or sneaky villain. The "skunk" or "polecat," they said with scorn and contempt.

And, like the old cowhand put it: "Well, them philosophers say 'a rose is a rose is a rose' but us that know him, say 'a polecat is a polecat is a polecat!' "

## WATER DOG

West Texas settlers had neither time or inclination to ponder much about the scientific aspects of this region.

But in the sunsets of their lives, some pioneers did, and do, reflect and wonder especially about the animal life.

"Take that old 'water dog' for instance," mused an oldtimer after recent rains that are sure to bring out this tailed amphibian in good numbers.

"He's slick, slimy and looks repulsive."

"Not worth a dern, except for catfish bait," said the oldtimer.

"Water dogs" (really salamanders) live in muddy, stagnant tanks, lakes, ponds, wells, cisterns, cellars—always a damp place. And when it rained a great deal out here in pioneer days—that meant dugouts.

If a body of water dries up, "water dogs" disappear. And reappear after heavy rains.

* * *

Three hundred fifty million years ago, when waters were teeming with life, and plants were beginning to flourish on the land, a fishlike creature left the sea and crawled ashore, where it prospered.

But there was one way of life this creature could not change. Each spring, it had to return to lay eggs so that its young might be cradled by the gentle sea. For this cold-blooded animal was amphibian—living half in water and half on land.

A wide variety of these amphibians dominated the swamps for almost 100 million years, then died away, until today, but three groups still are alive—frogs and toads; salamanders and newts; and legless, snake-like blind worms.

The lizard-shaped salamander lives in many parts of the world, and it was said in ancient folklore to be immune to fire.

Some have lungs (water dogs); some have gills (mud puppies); and others breathe through their skin.

Spotted, with yellowish-hued coloring near the flat head, "water dogs" are not poisonous.

* * *

One time at Spur, when I was a boy, my dad, who fired an engine at the combination light plant and gin, told this story:

"After a big rain, water dogs came creeping out of the old Spade Tank, south of town, and crawled by the hundreds toward the warm engine room.

"I had to nail a 1x12 plank at the door bottom to keep 'em out. Before I could get it nailed, dozens of them were all around me in the engine room.

"I took a shovel and tossed the water dogs right into the coal-burning boiler furnace. That fixed them!"

"And it disproved," Pa said, "some old folklore of the dim and distant past that salamanders are immune to fire."

# # #

## THE BUFFALO GAP PAINTER

Coyotes were coyotes; wolves were wolves; but the "mountain-lion-like" panther was something else!

His high-pitched scream was like the scream of a woman in distress. He coughed and he growled! He was a crafty hunter and a fierce fighter.

Screams of a panther, sometimes called "painter," made cold chills run up and down the spines of pioneers, especially women and children.

Mrs. Cynthia Elizabeth Holt of Lamesa could tell about one that ranged from Buffalo Gap to the Caprock:

* * *

"My father," Mrs. Holt wrote: "moved out here from San Saba County in 1890. He bought some land and we settled.

"One time when I was six years old, my sister, Etta, and I were down at our water tank doing the wash.

"I wasn't large enough to be of much help. But could hlep some. Our tank was some distance from our house.

"We were just about finished washing clothes when we heard a panther scream—sounded just like a woman!

"My sister grabbed her arms full of clothes. I was just big enough to drag a chair full of clothes toward the house.

"We hurried as fast as we could. There was a good deal of brush to get through.

"We rushed to tell our father. He took our dog to where the sound came from, to pick up the panther's trail.

"Old Ring wouldn't do anything but bristle!

"We notified a neighbor, who had a pack of hounds, but by the time the hounds got there, the trail of the big cat was cold.

"They couldn't pick it up.

"Several years later, my youngest sister and I heard a panther scream along a wagon road through 'The Gap.'

"A neighbor and his daughter were returning from Abilene one night and they heard the panther scream, too!

"Quite a few others also heard it, in the same area we did."

* * *

The panther variously has been called mountain lion, cougar, puma and catamount.

# # #

## "WHITEY," THE MISSING LINK?

Time was out here in the West when you wanted an ant bed "cleaned out," you went out and found a horned frog to do the job.

Is he a horned frog or horned toad? Well, if you want to get technical about it—he's a horned lizard.

But to small lads, not afraid to pick him up, he is a "horny toad!"

A Texas law now protects him from extinction because thousands upon thousands of his kind were captured, traded or given away as a novel representative of the West. Too, he got to be highly commercialized.

For a good many years, West Texans have pondered the possibility that horned toads descended from giant reptiles that lounged around in a prehistoric age.

There has been little if any proof of that theory. But a remarkable find—that of a phantom-like, solid white horned frog—may renew the "missing link" theory or speculation.

C.A. McClain of Spur, wrote:

"In 1935, I, and two sons, Eldon and Jessie, broke, with several sticks of dynamite, a huge flat sandstone rock on my farm in northeast Garza County.

"The rock must have weighed 10 to 12 tons. It was even and flat on top and was about 2½ feet thick. It had about 100 square feet of surface.

"A dynamite charge broke the rock into eight or nine pieces. In one, there was a white horned frog in a cavity, shaped much as if it had been done with a spoon.

"The explosion scratched the toad's head and it bled. He was very much alive. We did no more than look at him. But I noted he was still alive three hours later. 'Whitey' had horns or spines.

"The rock had no sign of a crevice in it prior to the dynamite blast.

"The white horned frog was not at all like the brown kind we see in West Texas.

How did that horned frog get there? How long was it entombed? Amateur scientists, at least, claim horned frogs can live in a dormant state 15 to 25 years.

But this white one? Did the sands of time cover him? Did he bleach from lack of sunshine? What did happen?

# # #

# BR'ER COYOTE IN THE WATERMELON PATCH

Br'er Coyote is a cunning creature.

He is a rascal, sort of like Br'er Fox of Joel Chandler Harris' Uncle Remus stories.

Out here in West Texas and Eastern New Mexico in the early days, he could become a probelm to ranchers and farmers.

As the World Book Encyclopedia says, he was helpful in one way by eliminating rats and other rodents.

But in the winter months, he was not so helpful. He would turn to poultry and livestock, if he got hungry enough. And he usually did.

He is listed as a predator.

The coyote was not feared as much as his larger cousin, the wolf, but he was consistent and always seemed to be around on the prairies. His lonesome bark was much heard on the lonely outposts of civilization.

Now, old Br'er Coyote liked watermelon, and he was a crafty invader of many a patch.

He's still around in reduced numbers, of course, but he still likes watermelon.

Freddie Howard of Brownfield, a big producer and handler in the expanding watermelon commerce on the Plains, said coyotes still are a problem.

He contends, (and we've heard this was true as told by oldtimers), the coyote has an uncanny knack for selecting only ripe watermelons.

He bites right into them and feasts on the heart of the melon. His appetite for them is tremendous.

And he never picks out an unripe one, either.

So how does modern scientific watermelon growing cope with Brother Coyote?

No, it does not trap him. He seems to have a built-in ability to avoid snares and traps.

Nor do timed devices—guns and cannon—that emit gases or poisonous fumes, hustle him out.

Timed devices that go off at intervals with a loud bang and scare the daylights out of an old, mangy coyote, feasting on a ripe, juicy and plump watermelon, work best.

He takes off, fearing the grower has opened fire with a shotgun. That was one possibility said to deter many a boy, too, bent on "just borrowing" a big watermelon.

So there you have it—the modern way to beat Br'er Coyote, who has such a taste for sugary, juicy watermelons on a soft, summer night.

Ah science! Isn't it wonderful?

"Poppa" Dortch on Duck Creek, or the late Jim Cloud's dad on a sandy land slope of Soldiers' Mound, sure could have used some of it, way back then, and when.

# # #

## FEATHERS OF GOLD

In western Kent County of West Texas, along the Salt Fork of the Brazos River, near its confluence with Catfish Creek (White River), is a beautiful scenic spot called Eagle Bluff.

Here, on a high rocky pinnacle on the south bank of the river—the majestic Golden Eagle (Mexican Eagle) once built its nests.

This bird, one of the largest of eagles, had brown plumage but its tawny feathers on the head and neck shimmered like gold in the sunlight.

As late as the 1920s, one lone eagle returned to this spot, vainly searching for its mate (eagles mate for life).

Now he is gone. And the Golden Eagles no longer soar above the river bluff.

* * *

Since Eagle Bluff was in an area once enclosed by fences of the 24 Ranch, owned by Boley Brown and Pete Scoggins, there is a story:

Now Brown was a salty rancher and he loved good horses. He was a stickler for cowhands who would not endanger his horses.

One day in Spring, a cowhand, riding a 24-branded mount, happened to be checking water gaps on the river in the area of Eagle Bluff.

He was attracted by a Golden Eagle circling over the steep bluff, of all things with a rattlesnake in its talons.

Then, he saw the big bird fly directly into an overhang, and quickly return without its catch, to a pattern of tighter turns, designed to lure the cowhand away from the bluff.

Fascinated, the puncher rode closer and spotted a large nest built of sticks, hanging precariously beneath a ledge.

As he searched for a place to ascend the bluff to the nest, he noticed a second eagle, soaring overhead—a mate, no doubt.

Finding no way to get down to the nest, he dismounted and secured his lariat rope to the saddle. His horse, trained to stand still on command, held a solid stance, as the cowhand slowly went down the rope to the nest. Getting back up was a harder job. But he accomplished it and rode back to the ranch house, carrying two eggs he had found in the nest.

Brown met him at the corral gate, wanting to know why he had returned so early. The cowhand proudly showed the two eagle eggs to the boss.

Brown wanted to know how the cowhand got to the nest, and when told, snapped: "You're fired!"

The salty rancher did explain to the puncher, "You not only risked your own life, but that of a valuable horse. Horses and hands are too valuable to allow such risks!"

* * *

The chagrined cowhand turned in his time and left the 24 Ranch, but the eagles returned to the nest beneath the precipitous bluff.

Again and again, they went there, to raise their young and enjoy a wild, free life in the rugged ranchland below the Caprock, until the time when only one Golden Eagle soared above the bluff.

And then, he, too, was gone...

# # #

## OLD PAINT

Probably more than any other breed of horse, the Paint is identified with frontier America.

The Paint or Pinto was the mount of those well known bad guys—the warlike Indians.

Virtually any sketch or painting of the Plains Indian shows him astride a Paint pony.

It's a safe bet the Indians who skewered General Custer that fateful day on the Little Big Horn, did so from the backs of Paint horses.

Paints pre-date the Indian wars, however. Drawings in Egyptian tombs date them from as far back as 3,400 B.C.

The American Paint Horse Association says two of the 16 horses brought to North American by Cortez in 1519 were Paints.

After the Indian fights, some cowboys chose Paints, and so this breed played a role in the winning and settlement of the West.

Ballads of those early days reveal that Paints were held in some regard, to wit: "...Gonna saddle 'Old Paint for the last time and ride..."

Despite this early day popularity, the Paint or Pinto went into an eclipse from which he is just now emerging.

Solid color breeds held sway for many years. The Paint was left out in a far corner of the horse pasture, so to speak.

The American Paint Horse Association, as organized in 1962, has come on with a rush. It lists members in more than 40 states.

Paints do all right in competitions of the horse world, too. "Calamity Jane," a Paint mare, won $1,200 in open cutting in a 1969 event.

Paint horses follow the conformation lines of both Quarterhorse and Thoroughbred, depending on their breeding.

Paints can be obtained by crossing Paint mares with solid color stallions, or vice versa.

There are two varieties of Paint horses—the Tobiano and the Overo.

"Simplest way to tell the two apart," according to Ranchman publication, "is to look at their backs. No white from the horse's withers to its tail usually indicates it is an Overo."

"Tobianos outnumber Overos because their genes are dominant," veterinarians state.

Here's a good "kicker" about Paints:

"Well, it's a two-tone age for automobiles, why not horses?" nickered a Paint pony.

## TO TAME A MOCKINGBIRD

To tame a mockingbird is quite a feat, very unusual.

But this, Mr. and Mrs. Lamar Curry of Lubbock, did.

It all began on a June afternoon (1971), when Curry was shaking plums out of a tree in his backyard.

He shook the tree vigorously and failed to notice that plummeting down with the fruit from the top part of the tree, was a newly-hatched mockingbird.

However, Chipper, the Curry's toy poodle, set up such a bark that Curry investigated. He found a baby mockingbird, so young it was devoid of feathers, lying in ground ivy at the foot of the tree.

Curry immediately knew he had shaken the bird from a nest located high in the plum tree. He also knew that if he touched the bird to put it back into the nest, the parent birds would not accept it.

Therefore, he took the little bird into the house and thus started an interesting experiment of how to tame a mockingbird, or make a pet of one.

There was a lot of trial and error as the Curry's sought what to feed the trembling little bird. Water it would take out of a spoon.

"Let's try him on plums since he came from a plum tree," Curry told his wife. Sure enough, the little bird liked the fruit.

But his favorite food turned out to be canned dog food—just the brand that Chipper ate.

The Currys made their pet a nest in an old pail hanging in the store room just off the backyard patio. Later, a little swing was added. From the first, the young bird liked to swing on this perch. Once, he was left swinging on a limb of the tree, but sparrows attacked him.

He hopped about inside the house and finally began to perch on the shoulders of either Curry or Mrs. Curry.

The young bird was fed six times a day. He was taken outside each day and learned to hop about and hedge-hop to try out his feathers, which he had grown.

The little mockingbird was left with a "bird-sitting" family next door when the Currys went out of town for a day. And he stayed put.

But the Currys knew that soon he would be flying away. They were not anxious for that day to come. They planned to band him and see if he returned to the plum tree next summer.

Alas! One of the cats in the neighborhood got him and that was the end of the story.

It is a familiar story in the city. Perhaps that is why the sweetest singing mockingbirds are far out in the country, away from urban living.

# # #

## OPOSSUM

Now and then, the opossum, or "possum," as he was called by most, showed up out here in West Texas as it was being settled.

Related to the kangaroo, the opossum has a furry body and rather ratlike face. Like other marsupials, the females have an external pouch in the skin of the abdomen. The young are sheltered in it. Opossums are the only marsupials native to North America.

The opossum we know in the South is larger than a cat when full grown. It has coarse, grayish white hair, a piglike snout, naked ears, and a long hairless tail, that is scaly. It can hang by the tail from tree branches and rafters.

Possums prowl at night. At a quick glance in semi-darkness, they look in the face somewhat like a small, graying man with a wizened face.

When they are in danger, they lie motionless and pretend they are dead. From this habit, we have taken the expression "playing possum!"

* * *

Phantoms, then, do not necessarily have to be a shaggy buffalo, howling wolf, screaming panther, or a wild stallion.

Like the one told about by Mrs. C.L. Simmons of Plainview:

"...A friend told me of her early-day experience with a phantom.

"Mrs. Graham's husband died early in life, leaving her childless but owner of a small cattle ranch, where she continued to live alone after his death.

"She was a brave, industrious woman.

"Mrs. Graham had about an acre or more fenced in around her house, where she kept young calves for protection from wolves and coyotes.

"One night, Jack, the big watchdog on a leash in the yard, kept growling.

"And 'frisky,' the house dog, was 'whiney' and restless. He kept running back and forth into a bedroom.

"Finally, Mrs. Graham took a lamp into the bedroom to find the cause of the dog's restlessness.

"On entering the bedroom, she glanced quickly at the window, and saw what appeared to be an old, gray-bearded man with piercing eyes and pointed nose, looking at her through the vines.

"Not wanting 'him' to know she had seen 'him,' she moved around the room a bit, then blew out the light.

"Going very quietly into the kitchen, she picked up an iron poker, slipped noiselessly out of the door, and unleashed Jack.

"Together, they headed for the window at which she had seen 'the peering old man.'

"She gave the 'old man' a wham on the head. She knocked him off a trellis.

"To her consternation, the 'phantom' turned out to be an opossum that had climbed the trellis and was peering in the window!

"This whiskered, furry fellow wasn't 'playing possum,' either—he was knocked out cold!"

# # #

## THE IMPROVISED THANKSGIVING

Time was growing short that fall in 1880...

Lizzie Campbell was worried. The festive holidays were approaching. Thanksgiving would be first, then Christmas.

A supply wagon had been sent all the way to Fort Worth to bring in food supplies to the remote Matador Ranch.

Somewhere, it carried a precious cargo of tame **turkeys**, cranberries, and other food for the holidays.

It was overdue...

Mrs. Campbell, official hostess of the Matador Ranch and wife of the founder of that vast spread, had to improvise—as so many cooks had to do in pioneer days.

Wild turkeys were available on Dutchman's Creek, and around Ballard Springs, where the ranch headquarters nestled.

But wild turkeys sometime were "gamy" and tough.

But Mrs. Campbell had an eye on two hogs that had been turned loose to fatten on acorns.

These large, wild hogs, butchered, would furnish strips of fat to be laid on the stringy turkeys for seasoning and cooking.

Mrs. Campbell prepared hams, venison steaks, and antelope stew with dumplings—called "sinkers" by the cowboys.

As a substitute for cranberries, she opened jars of wild plum jelly. She prepared wild rice and canned corn.

She baked apple pies, using dried apples, of course.

She prepared a tub of doughnuts. For afternoon snacks, there were popcorn balls and pecans.

What a festive board the ranch family and cowhands sat down to that early Thanksgiving.

And what's more, buckets of food were sent to cowboys at the line camps.

* * *

Wild turkey was a welcome change of diet from antelope and beef that ranchers and cowmen could enjoy.

The birds were plentiful along cottonwood trees, dotting creeks.

But not any more.

* * *

**There were so many wild turkeys in some places; early cowmen got tired of eating them.**

The late William L. (Uncle Bill) Hyatt of Spur once said in an interview: "I came to the Staked Plains in 1878 with a herd belonging to the '22' outfit (ranch). We followed the Mackenzie Trail right up to the Caprock, and turned Hensley longhorns loose on what later became land of the Matador and Spur Ranches, two of the 'big ones.'

"There was no one in the area when we came, excepting a few renegade Indians and buffalo hunters.

"The cottonwoods along the draws were full of wild turkeys, and we soon tired of eating them. A cowman prefers beef, you know. Our boss frowned on too much beef from the herd, so we ate buffalo calves..."

# # #

## A WONDERFUL BIRD IS THE PELICAN...

*"A wonderful bird is the pelican...*
*His beak holds more than his belly can..."*

So goes an old rhyme.

And that is what RC. Davis of Earth found out in 1966.

Family members and others went out seining for fish to feed a large white pelican that flew down onto the Plains.

Actually, the 16-pound bird landed on a farm near Lazbuddie, and later was captured by Davis, who also has a farm in that area.

David discovered the gawky, big bird in the afternoon and went back at night to get him. He blinded the bird with a flashlight and grabbed it, receiving a ripped finger in the process.

David hardly knew what to do with the bird, very uncommon out here on the Plains.

One thing for sure, he soon had to do something because Mr. Pelican almost ate him out of house and home. And every tank for miles around had to be fished to supply the pelican's whopping appetite.

The pelican stayed in a barn on the Davis farm. It had a wingspan from 8 to 10 feet, and flopped around quite a bit without a "word," excuse it—without a grunt, groan, peep or squeak, or whatever pelicans utter. They are said to be the "most silent" of birds.

* * *

The capture of a pelican near Lazbuddie by Davis in 1966, stirred the memory of a Pep community pioneer, A.G. Jungman, a taxidermist.

He recalled a pair that apparently were forcelanded from hunger in the area between Littlefield and Amherst on Blackwater Draw in 1946.

A farmer, on whose land they let down, had his troubles with the fish-hungry birds that threatened to catch all the fish from his personal tank, which had been well-stocked.

After chasing the two big birds away from the tank, time-after-time, with a shotgun, the farmer decided to shoot one, and hope the other one left.

He did. But the other pelican wouldn't leave.

So, the farmer brought a pair of pelicans to Jungman's taxidermy shop to be mounted. But they never were called for.

Jungman stuffed the birds and put them on exhibit at his shop. His grandchildren had lots of fun with the "funny birds" that were mounted, according to the Pep pioneer.

Jungman also recalled that, one time, a white pelican was captured near Levelland and it stayed for "quite awhile" on the City Park Lake there.

Jungman adds a line to the old pelican rhyme:

*"A wonderful bird is the pelican...*
*His beak holds more than his belly can..."*
*"I don't see how the heck it can!"*

# # #

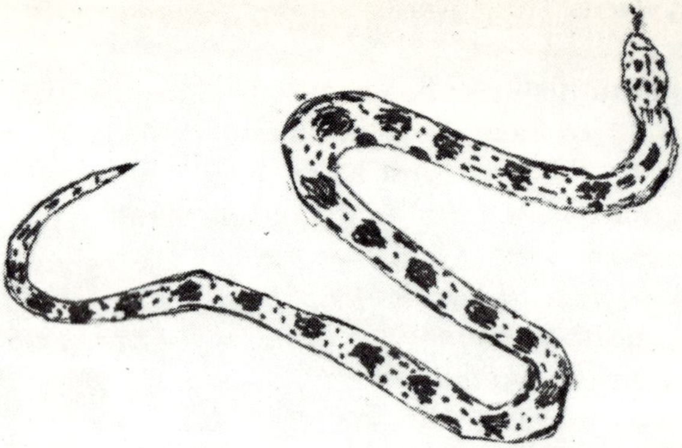

## NEST ROBBER

It wasn't long after turn of the century that Anderson and Margaret (Maggie) Wood and their son left Fisher County to go farther "out West" to Kent County up close to the Caprock.

Wood's pick for a homestead was three sections of land, cut off from the old 24 Ranch, a few miles northwest of Clairemont.

The choice was perfect. It was situated on an "S" bend of the Salt Fork of the Brazos River, just south of the Red settlement of Dickens County.

There were hardships getting started. But Anderson and Maggie Wood were of sturdy stock from two families of Texas' earliest settlers.

While Anderson hitched up two huge work horses—old "Barney" and "Seelum"—and got to breaking out farm land to augment cattle grazing in the deep grass by the river, Maggie was raising turkeys and chickens.

Raising poultry was no easy task. Anderson built Maggie a "varmint-proof" chicken house. Nevertheless, threats to the poultry still existed.

There were large hawks, coyotes, bobcats and even eagles, around.

But an insidious phantom played the neatest trick of all on Maggie. It was like this (as Mrs. Wood told it):

* * *

She had a prize laying hen—good for a fine egg, day in and day out.

Suddenly, Maggie began to find an empty nest of straw under the prize hen when she went to gather eggs.

This went on for days. Mrs. Wood was puzzled. She decided to keep a vigil at the chicken house and find out what was going on.

Well, the old hen hopped up on the nest compartment, sat awhile, and then hopped to the ground below the row of nests, cackling like all get out.

A nice round white egg loomed in the straw nest.

Mrs. Wood started to retrieve it.

Suddenly, the straw parted and a large bull snake poked out, intent on getting his daily meal.

Certain it was no rattler, Maggie grabbed a stick and chased Mr. Bullsnake, scolding him sharply.

"Get going, you old bull snake, you, I've found you out at last!"

Ordinarily, she would have let the non-poisonous snake go, but not this time—she disposed of him hurriedly with a hoe.

And Maggie Wood continued to get an egg a day from her prize laying hen.

# # #

## PRAIRIE CHICKEN

We are saying goodbye to another friend of the prairies.

Like the buffalo, antelope and prairie dog, the prairie chicken—once a succulent morsel on the prairie pioneer dinner table—is vanishing.

The bird now survives only in small, isolated colonies where once it abounded on the wide plains and prairies in goodly numbers.

The decline is due not to hunting, against which the law has protected the prairie chicken for many years, but rather to habitat destruction.

The tall grass and shrub areas, once a cattle grazing paradise, have become productive farmland.

The prairie chicken was here in great numbers, all right. And had his day. What did he do? How did he live? What were his characteristics?

For one thing and common in the grouse family, there was a ritualized mating episode. It took place in early spring on what was called a "booming ground." Here, the males "strutted their stuff" in a sort of wing-beating combat for nesting territories.

Their short tails were widely fanned out like a peacock's. And air sacs on each side of the neck were inflated to distortion. A booming "coo-hooooo" call was sounded.

* * *

This courtship of the prairie chicken, a fascinating dawn ceremony, has been under study by Eastern New Mexico University students at Portales.

Biology majors shiver in the pre-dawn as they go out to shinnery-covered pastures of southern Roosevelt County, about 15 miles south of Elida.

An intensive study is being done in cooperation with the New Mexico Game & Fish Dept. The specific aim being to gain information on the mortality rate of the young prairie chickens.

The study is coordinated by Dr. A.L. Gennaro, whose speciality is ecology.

The mystery of the booming grounds is fascinating. Why, researchers wonder, do the birds come back to the same area each spring?

The students use nets to catch the birds for examination and banding.

A recent innovation is to "bug" the birds with tiny transmitters, attached with harness so as not to interfere with flight, feeding, or other activities.

The idea is for weak transmission signals to guide study teams to nesting sites.

There's a hitch, though, to this scientific study—only the males frequent and use the booming grounds.

And, there is little roaming from one booming ground to another. Of more than 500 birds tagged, only three have shown up in more than one booming ground.

But one old male, his feathers showing wear and tear, was captured four times in one week in the same net, as the catchers moved from one grounds to another.

One, perhaps, more aware than his fellows of the necessity for doing something about the declining prairie chicken population.

# # #

## MOONLIGHT AND RACCOONS

Workmen on a nearly-completed dormitory on the campus of Texas Tech one January morning in 1963, were puzzled by what appeared to be tracks of a small child in dust on the floors of the building.

"Those are coon tracks," said Larry Estep, office manager for contractor H.A. Lott, Inc., Houston.

The workmen began leaving apples, other fruit and nuts about the construction site, and each morning these food items had disappeared.

But near a water barrel, peelings and nut shells were found—evidence that a coon had followed the age-old practice of his kind, to wash all food before consuming it.

Its eyes shining, one day workmen did capture a raccoon

in the dark, work elevator shaft. They kept and fed the animal until completion of the structure.

* * *

Raccoons have been out here in West Texas a long, long time. "The old raccoon and light of the moon," have furnished hunting thrills as many years.

And these "masked" phantoms of the night have reappeared close in to "civilization" from time to time.

* * *

A South Plains Coon Hunters Association was formed in 1955, with 10 charter members. Membership grew to 64 by 1958.

Monthly meetings would be held on the old Johnson Ranch in Yellowhouse Canyon near Lubbock. There would be as many as 70 coon dogs on hand, all barking at the same time but in different voice.

Activities would include competitive events of "Treein'," "Waltzing the Keg," and "Coon in the Log."

Some of the standout performers included: "Red Head," owned by I.B. Roberts, Sweetwater; "Tip," fine coon dog of Don Faulkenberg, Rt. 2, Floydada; "Rock," entered by J.H. Davis of Hart; and others.

* * *

There is something that gets in the blood of a coon hunter. He develops ears attuned to the calls of his dogs. He calls it "sweet music." He can chart the progress of the hunt by the baying of the coon dogs, far afield. He hurries up to where the coon is "treed" at just the right time. Many call it "great sport."

* * *

The raccoon is called "coon" for short. It is related to the bear. The common raccoon is about 32 to 34 inches long, from nose to tip of tail. It weights from 20 to 25 pounds. The stout body is covered with long, coarse hair. The hair is grayish with black tips and "tail rings." The face is like that of a fox. A black patch at each eye, gives a cunning, roguish, masked look. Raccoons are famous tree climbers.

In earlier days of our country—coonskin caps (a la Davey Crockett and Daniel Boone), sleigh robes, and overcoats, were popular.

At one time, coonskin coats were the rage among college boys.

# # #

## "...AND THE ANTELOPE PLAY"

Ask any oldtimer who came to West Texas before turn of the century what animal life was here, and 10 to 1, he or she will mention antelope.

This prairie country was teeming with them. But here, because of lack of cover, the antelope became easy prey, and almost was extinct by 1900.

A continual closed season on the animal, also known as the pronghorn deer, here along the Llano Estacado, helps preserve token herds of the graceful, nimble creatures with large, innocent eyes.

There is a herd, browsing today at the western lip and in Yellowhouse Canyon, on the V8 Ranch, southeast of Lubbock. Often good grazing beckons the animals to the very city limits.

Here, they are seen sometimes munching contently within sensitive earshot of busy U.S. Hwy. 84, between Lubbock and Slaton.

The hustle and bustle of traffic, the sounds of industrial southeast Lubbock, the metallic clip-clap of passing freight trains, even screaming jets from Reese Air Force Base or Lubbock's busy airport, do not stampede them.

They can be approached in a pickup truck or automobile. But if the motorist steps out of the conveyance—look out! They are gone in a wisp of brown and white in kangaroo-sized jumps.

You will know Mr. Antelope if you see him. His upper parts—rump, tail, "face sides," and chin—are white.

Two white bands extend under the throat. The mane is russet and tipped black.

In the early 1970s, the V8 Ranch had a herd of about 32 animals. But the toll to predators and "poor sportsmen," is high. Sometimes, wild dogs from the city, in packs, attack.

The V8 herd has ranged the five-section spread since the 1940s. The original stocking was about 10 does and one buck. The herd has gone as high as 37 animals.

Quite a few of the area ranches, including some of the big ones at the edge of the Caprock, also have token and pet herds of antelope.

An incident of 1890 will illustrate the large number of antelope around Lubbock:

During a prolonged snow, cattle and antelope together drifted southward to the fence of the IOA Ranch near Singer's Store on the Yellowhouse.

IOA hands rode out to drive the cattle back to sheltering canyons. They found 3,000 antelope with about 5,000 head of cattle.

The antelope were "snow blind." As the snow melted, the antelopes' blindness cleared and the large herd disappeared.

Many stories are told of their antics and how plentiful antelope were in West Texas.

And many persons today, watch to see the "little fellers," in the beautiful spot of Yellowhouse Canyon near Lubbock, "where the antelope still play."

# # #

## "P-TRACK! P-TRACK! P-TRACK!"

How long has it been since you have seen a "guinea" or heard his harsh, monotonous cry of "p-track, p-track?"

At one time, guinea fowl were quite common out here and almost all farmsteads and ranches had flocks of them.

Because the guinea fowl made such a noise when disturbed, farmers and ranchers valued them as "watchdogs." At the first hint of danger, the guinea fowl's harsh, strident clamor alerted farmer or rancher of a possible predator, be it human or threatening hawk, slinking cat or weasel.

Their ear-splitting cackle—released at the slightest provocation—could be a nuisance around the farm or ranch house, too.

"There were lots of guineas around the ranch homes down below the Caprock," said V.L. (Dogie) Graves of 2208 24th St., Lubbock. "They sure could spot those old chicken hawks," the cattleman said.

"Haven't seen or heard one in years," said Jimmy Yates of 3012 22nd St., Lubbock, "but I remember one time when I ran over one with a car and killed it and made the mistake of stopping and asking at a farm house if they wanted the guinea to eat and I would pay damages," Yates added.

"That farmer's wife came after me with a broom and she made a lot more noise than any flock of guineas ever did," Yates laughed.

Avalanche-Journal reporter Leland Boyd's father, a Central Texas farmer who always raised some guineas, was named Jack Boyd. Son Leland said the neighbors teased his dad saying those old guineas were cackling: "Pa-Jack! Pa-Jack!"

In size between a chicken and a turkey, the guinea fowl roosts in trees in preference to poultry houses. Its eggs are good and the flesh is considered a delicacy. Guineas are hard to raise and have a bad habit of hiding nests. The fowl congregates in flocks and destroy many insects.

Guinea fowl are found in the wild only in Africa and Madagascar. All seven species have bare heads and necks, sleekly curved bodies of smooth feathers. The ones we have in West Texas have gray plumage with white specks or spangles.

The guinea fowl apparently was domesticated by the ancient Greeks and Romans, who brought them from Africa. The bird apparently was unknown to medieval Europeans until the intrepid Portuguese navigators of the 15th Century ventured down the coast of Africa, and brought some of them back from the Guinea coast.

The old guinea never made it as an egg producer. Most raised today are for the gourmet's market.

Though essentially a ground bird, the guinea fowl can fly strongly. There have been attempts to introduce guineas into New World prairies and brushland as a game bird, but its habit of running from danger, not flying, does not make for good shooting sport.

Guinea fowl are especially noisy when going to roost, and one oldtimer said they "sounded like a bunch of teen-age girls in a dormitory at bedtime!"

If you get homesick to see or hear them, some still are around. At least one Lubbock resident has some. Mrs. Clarence Gregory reports about 150 guinea hens on their farm near Ralls, a project that started with a hobby. The James farm on FMR 1585, a half mile west of the Tahoka highway near Lubbock, has a small flock of about eight guineas. "People stop and look at them and some buy them," said Mrs. James. "I like to hear that old 'pot-rack, pot-rack'," said Mrs. Leona Lamb of Slaton. Her son Wayne Lamb, raises guineas, at his place near Slaton.

# # #

*Designed by*

Neil C. Vanzant